Springer Undergraduate Mathematics Series

The Springer Undergraduate Mathematics Series (SUMS) is a series designed for undergraduates in mathematics and the sciences worldwide. From core foundational material to final year topics, SUMS books take a fresh and modern approach. Textual explanations are supported by a wealth of examples, problems and fully-worked solutions, with particular attention paid to universal areas of difficulty. These practical and concise texts are designed for a one- or two-semester course but the self-study approach makes them ideal for independent use.

Paul C. Matthews

Differential Equations, Bifurcations and Chaos

 Springer

Paul C. Matthews 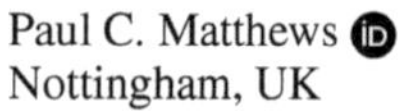
Nottingham, UK

ISSN 1615-2085 ISSN 2197-4144 (electronic)
Springer Undergraduate Mathematics Series
ISBN 978-3-031-99542-2 ISBN 978-3-031-99543-9 (eBook)
https://doi.org/10.1007/978-3-031-99543-9

Mathematics Subject Classification: 37-01, 34-01

Preface

This book introduces the topics of nonlinear differential equations, bifurcation theory and chaos. Some of these concepts are relatively recent in the history of mathematics, yet can be understood without very advanced mathematical knowledge. In fact, much of the material involves describing and sketching solutions rather than finding exact mathematical formulas for them. This field of nonlinear dynamical systems has a wide range of applications across the natural sciences.

When I first started learning about this subject in the 1980s, it was usually taught as postgraduate level material, or as a specialist third-year undergraduate option. But these days, many undergraduate mathematics courses include nonlinear differential equations and chaos in the second year, and some of the material is often taught in the first year. Hence, many of the excellent older textbooks on the subject are set at rather too high a level compared with the stage at which students encounter this topic. This book is based on many years' experience of teaching differential equations to undergraduate students.

The book is aimed at students studying mathematics, natural sciences or engineering. It is assumed that the reader is familiar with calculus, including Taylor series and partial differentiation, and linear algebra, including vectors, matrices, eigenvalues and eigenvectors. An Appendix is provided, giving brief summaries of some of these key topics. Some experience of computer programming will also be very helpful, particularly as an aid in understanding the topic of chaos in the final chapter.

As well as worked examples throughout the text, there are exercises at the end of each chapter that form an essential part of the book. This subject is best learnt by doing rather than reading. Solutions to the exercises are given, but you will learn more if you attempt the exercises without looking at the solutions. Some of the solutions are intentionally brief, leaving the reader to fill in the details. Many of the exercises are similar to the worked examples.

This book is written in the style of UK applied mathematics. The emphasis is on methods and understanding rather than precise mathematical rigour. There are a few stated theorems, but no formal proofs, so those who enjoy books that consist mainly of alternating theorems and proofs will be disappointed.

The book is deliberately concise, in keeping with the Springer SUMS style. The aim is to lead the reader through the mathematics necessary for an appreciation of the topic of chaos in the final chapter. So although the equations are mainly nonlinear, the narrative is linear, with each new topic building on the previous ones. Inevitably, this means that some interesting aspects of nonlinear systems are omitted. A list of recommended further reading is provided at the end of the book.

I am grateful to the many people who have helped me understand this fascinating topic over the years. These include Paul Glendinning, Rebecca Hoyle, Willem Malkus, Michael Proctor, Alastair Rucklidge, Steven Strogatz and Nigel Weiss.

I would also like to thank Andrew Gilbert, Rachel Nicks and Alastair Rucklidge for their careful proofreading, and everyone at Springer who was involved in the initiation, planning and production of this book. The text of this book was written using Tex Live and TeXworks, and the diagrams were produced using GNU Octave.

Nottingham, UK   Paul Matthews
May 2025

Declarations

Competing Interests The author has no competing interests to declare that are relevant to the content of this manuscript.

Contents

Chapter 1
Introduction

Some people say that money makes the world go round. Others say that love is the driving force. But both of these are wrong. The saying should be:

Differential equations make the world go round.

The Earth is pulled towards the Sun by the force of gravity. Sir Isaac Newton's famous second law of motion,

$$\text{force} = \text{mass} \times \text{acceleration},$$

is in fact a differential equation, because the Earth's acceleration is the rate of change of its velocity. The solution of this differential equation is that the Earth goes around the Sun in an elliptical orbit, as observed by Kepler and then proved mathematically by Newton. Similarly, the Earth rotates on its axis, slowed down slightly by friction due to the tides, leading to a differential equation. This in turn leads to a slow exponential decrease of the rotation rate, so that each day is a fraction of a microsecond longer than the previous one.

By a similar argument, any problem in mechanics or dynamics leads to a differential equation, via Newton's law. If you want to predict the motion of a swinging pendulum, or calculate when a falling ball will hit the ground, or predict when a comet will return, or find the rate of water flow out of a reservoir, or predict the weather, you need to solve a differential equation. On the very small scale, particles behave like waves, and their motion is described by the Schrödinger equation of quantum mechanics, which is a differential equation. At the other extreme, the large-scale structure of the universe and the curvature of space-time due to the presence of matter are determined by Einstein's field equations, which are differential equations. The fundamental equations describing electric fields, magnetic fields and light waves are Maxwell's equations, which are differential equations.

© The Author(s), under exclusive license to Springer Nature Switzerland AG 2025

P. C. Matthews, *Differential Equations, Bifurcations and Chaos*,

Springer Undergraduate Mathematics Series,

https://doi.org/10.1007/978-3-031-99543-9_1

In addition to describing mechanics and motion, differential equations are widely used to model all sorts of practical problems. Anything that involves a quantity that changes with time or space can be described by differential equations. Examples include chemical reactions, radioactive decay, population dynamics and the competition between different species, the wave of infection during an epidemic, the growth of tumours, and even traffic jams.

But there is a problem. Differential equations are hard to solve. In fact, only a few very simple types of differential equation can be solved using analytical methods. The most important of these methods are described in Chap. 2. Faced with this difficulty, there are various possible approaches. One is to use a numerical method on a computer to obtain an approximate solution. Or, if the differential equation has a small parameter, and the solution is known when that parameter is zero, then an approximate solution can be found using an expansion in powers of the small parameter. A third approach, which is the main theme of this book, is to use *qualitative methods* to describe the solution. This means that although we cannot say exactly what the answer to a problem is at some specific time, we may be able to describe and sketch the solution, and say, for example, that the solution oscillates and tends to zero as time increases.

Qualitative methods for first-order differential equations are described in Chap. 3, and Chaps. 4 and 5 consider the second-order case. Chapter 6 introduces the idea of a bifurcation—a change in the behaviour of an equation that occurs as a parameter in the system is varied. The discrete analogs of differential equations, known as difference equations, are discussed in Chap. 7. Finally, the relatively new subject of chaos is introduced in Chap. 8. Apparently simple differential equations and difference equations can have remarkably complex behaviour, showing irregular oscillations that are very difficult to predict.

Chapter 2
Analytical Methods for Differential Equations

This chapter gives some important definitions regarding differential equations, and describes how they are classified into different types. Standard analytical methods for solving certain classes of differential equations are introduced. These include first-order separable equations, first-order linear equations, and second-order linear equations with constant coefficients. These methods are not the main theme of this book, but are essential background material that will be needed for the later chapters.

2.1 Definitions, Classification and Notation

This section introduces some key definitions regarding differential equations, describes how they are classified into different types, and explains some common notation that will be used throughout the book.

2.1.1 Differential Equations

A *differential equation* is an equation involving one or more derivatives of a function. Suppose that the function is $y(x)$. Then the variable x is known as the *independent variable*, and y, which depends on x, is the *dependent variable*. To save some writing, we will usually just write y rather than $y(x)$ when it is clear from the context that y is a function of x. The choice of symbols used for the variables depends on the context. For example, in a system that evolves in time, t is often used as the independent variable. The variable x is commonly used as either the dependent or independent variable, as seen in the following two examples.

© The Author(s), under exclusive license to Springer Nature Switzerland AG 2025
P. C. Matthews, *Differential Equations, Bifurcations and Chaos*,
Springer Undergraduate Mathematics Series,
https://doi.org/10.1007/978-3-031-99543-9_2

Example 2.1 Suppose that $y = x^2 + 2x$. Differentiating y shows that

$$\frac{dy}{dx} = 2x + 2 \qquad \text{and} \qquad \frac{d^2y}{dx^2} = 2. \tag{2.1}$$

These are two differential equations that $y(x)$ obeys. It is easy to create more differential equations for the same function, for example

$$\frac{dy}{dx} = \frac{2y}{x} - 2 \qquad \text{and} \qquad \frac{d^2y}{dx^2} = \frac{dy}{dx} - 2x. \tag{2.2}$$

However, usually the reverse process is required: we have a differential equation, perhaps resulting from some physical law, and we want to find a function that obeys that equation.

Example 2.2 Suppose that a particle of mass M experiences a force F that varies with time as $F(t) = \sin t$. If the position of the particle is $x(t)$, then using Newton's second law, force $=$ mass $\times$ acceleration, $x(t)$ obeys the differential equation

$$M\frac{d^2x}{dt^2} = \sin t. \tag{2.3}$$

There are two abbreviations that are often used to write derivatives. If the independent variable is a space variable, x, the derivative of y with respect to x is often written as y', so the two differential equations in (2.1) can be written as $y' = 2x + 2$ and $y'' = 2$.

If the independent variable is time, then a dot over the dependent variable can be used for each derivative, so (2.3) can be written as $M\ddot{x} = \sin t$.

In many physical systems, the independent variable does not appear explicitly in the differential equation. For example, if the force on a particle depends on its position and velocity but not on time, its equation of motion is $M\ddot{x} = F(x, \dot{x})$. Differential equations with this property are said to be *autonomous*.

2.1.2   Order

The *order* of a differential equation is the highest derivative occurring in the equation. So, for example,

$$\frac{dy}{dx} + y^2 = x + xy \tag{2.4}$$

is a first-order differential equation for the function $y(x)$. Similarly, (2.3) is second-order, and

$$y''' = 2y'' + x^2 y \tag{2.5}$$

is a third-order differential equation.

2.1.3   Linearity and Nonlinearity

Differential equations are either *linear* or *nonlinear*. The equation is linear if each term is a linear function of the dependent variable or one of its derivatives. So, for a function $y(x)$, a linear equation can contain terms involving y multiplied by any constant or any function of x, and similarly for the derivatives y', y'', ..., but it cannot have any nonlinear functions or products of y and its derivatives, such as y^2, $\cos(y)$, yy'', and so on. If an equation has any nonlinear terms such as these, it is *nonlinear*.

The linearity or nonlinearity of a differential equation relates to the *dependent* variable. How the independent variable appears in the equation does not matter.

Of the two equations in the previous section, (2.4) is nonlinear, because of the y^2 term, and (2.5) is linear.

Example 2.3 The differential equations

$$x\frac{dy}{dx} + x^2 y = e^x y \qquad \text{and} \qquad \frac{d^3 u}{ds^3} + \frac{du}{ds} = s^3 \tag{2.6}$$

are both linear. But

$$x\frac{dy}{dx} + xy^2 = e^x \qquad \text{and} \qquad \frac{d^3 u}{ds^3} + u\frac{du}{ds} = s^3 \tag{2.7}$$

are both nonlinear.

The distinction between linear and nonlinear differential equations is important. Linear differential equations are often easy to solve, and they tend to have fairly simple behaviour. Nonlinear ones usually cannot be solved analytically, except for the simplest types, and tend to have more complicated and interesting behaviour, as we shall see towards the end of the book.

2.1.4   Homogeneous Linear Equations and the Superposition Principle

There is one more key definition that only applies to linear differential equations. If each term in the equation includes the dependent variable or one of its derivatives exactly once, the equation is *homogeneous*. But if there is a term that does not include the dependent variable, for example a constant term or a function of the independent variable, the equation is *inhomogeneous*. For example, of the two linear differential equations in (2.6), the first one is homogeneous, but the second is inhomogeneous because of the s^3 term.

An important property of any linear, homogeneous differential equation is the *superposition principle*: if $y_1(x)$ and $y_2(x)$ are solutions of the equation, then any linear combination of y_1 and y_2, $Ay_1(x)+By_2(x)$ is also a solution, for any constants A and B. This result follows directly from the linearity of differentiation.

For example, consider the simple second-order differential equation $y'' = 0$, which is linear and homogeneous. $y = x$ and $y = 1$ are both solutions, so by the superposition principle, $y = Ax + B$ is also a solution, for any A and B.

However, for the inhomogeneous equation $y'' = 2$, valid solutions include $y = x^2$ and $y = x^2 + x$, but a general linear combination of these two solutions is not a solution of the equation.

2.1.5   Constants, Boundary Conditions and Initial Conditions

Consider what is probably the simplest possible differential equation,

$$\frac{dy}{dx} = 0.$$

This equation tells us that the slope of the function $y(x)$ is zero, so y does not vary with x and hence y is a constant. The solution is $y = C$, where C is any constant. We can think of C as a constant of integration.

For the equation

$$\frac{d^2y}{dx^2} = 0, \tag{2.8}$$

we can integrate both sides with respect to x to get

$$\frac{dy}{dx} = C,$$

and then integrate again to find the solution

$$y = Cx + D,$$

where C and D are any constants. This solution including two arbitrary constants is known as the *general solution* of (2.8).

These very simple examples illustrate an important feature of differential equations. The general solution of a first-order differential equation includes an arbitrary constant, and the general solution of a second-order differential equation includes two arbitrary constants. If the differential equation is nth-order, its general solution should have n arbitrary constants.

A solution including arbitrary constants is not useful if we want to use the differential equation to predict something about a real-world problem. Some extra information is needed to find the values of the constants and hence find a specific solution that we can evaluate.

This is what we should expect from common sense. Suppose we want a mathematical model to describe how the concentration of some chemical changes during the course of an experiment. To do this we need two things: a rule that tells us how the concentration changes with time, which is the differential equation, and we also need to know the concentration of the chemical at the start of the experiment.

In the case where the independent variable is time, the pieces of information needed to find the constants are called *initial conditions*. For a first-order equation describing chemical concentration, this would be the initial chemical concentration at the start of the experiment. For a problem involving the motion of a particle, such as (2.3), which is second-order, two pieces of information are needed to find the two constants of integration, which would normally be the position and the velocity of the particle at the initial time. The combination of the differential equation and the necessary number of initial conditions forms what is called an *initial value problem*.

If the independent variable represents space, the extra conditions that determine the constants are called *boundary conditions*. An example of this is when water flows along a channel between two boundaries. At each boundary, the water is in contact with the boundary so its velocity should be zero. This gives two boundary conditions, which enable the second-order differential equation for the flow of the water to be solved. The term *boundary value problem* is used to describe the differential equation combined with the necessary number of boundary conditions.[1]

We will use the term *specific solution* to describe the solution of an initial value problem or a boundary value problem. That is, a solution of the differential equation that contains no unknown constants and obeys the initial or boundary conditions.

Example 2.4 Find the general solution of

$$\frac{dy}{dt} = \sin t, \tag{2.9}$$

and find the specific solution that obeys the initial condition $y(0) = 0$.

Integrating both sides of the equation gives the general solution $y(t) = -\cos t + C$, where C is an arbitrary constant. To obey the initial condition $y(0) = 0$, we set

---

[1] The questions of existence and uniqueness of solutions to differential equations is beyond the scope of this book. In fact, boundary value problems in particular may have more than one solution, even infinitely many. This book will be mainly concerned with initial value problems.

$t = 0$ in the solution, giving $0 = -1 + C$, so $C = 1$ and the specific solution is $y = 1 - \cos t$.

Example 2.5 Find the general solution of

$$\frac{d^2 y}{dx^2} = 2, \tag{2.10}$$

and the specific solution that obeys the boundary conditions $y(0) = 0$, $y(1) = 4$.

Integrating twice gives $y = x^2 + Cx + D$, where C and D are constants. This is the general solution. To find the specific solution, apply the boundary conditions. The condition $y(0) = 0$ gives $D = 0$, and $y(1) = 4$ gives $1 + C = 4$, so $C = 3$ and the specific solution is $y = x^2 + 3x$.

Remark 2.1 Having found the solution, it is a good idea to check that it obeys both the differential equation and the initial or boundary conditions. For the case of Example 2.5, the solution found was $y = x^2 + 3x$. By differentiating this equation twice, we get $y'' = 2$, which is the differential equation we started with. Also, substituting $x = 0$ gives $y = 0$, and setting $x = 1$ gives $y = 4$, so both the boundary conditions are obeyed. This completes the check.

The following sections of this chapter briefly describe some of the standard techniques for solving certain types of differential equation.

2.2 Separable First-Order Equations

A first-order differential equation for a function $y(x)$ is *separable* if it can be written in the form

$$\frac{dy}{dx} = f(x)g(y). \tag{2.11}$$

The derivative can be written as the product of a function of x and a function of y.

Example 2.6 The equation

$$x\frac{dy}{dx} = y^2 + 1$$

can be written as

$$\frac{dy}{dx} = \frac{1}{x} \times (y^2 + 1)$$

so it is separable. However,

$$x\frac{dy}{dx} = y^2 + x^2$$

is not separable, because it cannot be rearranged into the form (2.11).

The method for solving a separable equation (2.11) is to 'separate' the x terms to one side of the equation and the y terms to the other side, and then integrate, giving

$$\int \frac{1}{g(y)}\,dy = \int f(x)\,dx. \tag{2.12}$$

If we can do the two integrals, then we have found the general solution of the differential equation.[2] For each integral there is a constant of integration, but these constants can be combined together, so the solution involves one constant, as we expect for a first-order differential equation. This method is known as *separation of variables*. Since the method involves dividing by $g(y)$, we should first consider whether $g(y)$ can be zero.

Example 2.7 Find the general solution of

$$\frac{dy}{dx} = \alpha y, \tag{2.13}$$

where α is a constant.

The equation is separable, so we divide by y, noting that $y = 0$ is a valid solution of (2.13), and then integrate:

$$\int \frac{1}{y}\,dy = \int \alpha\,dx \;\Rightarrow\; \log|y| = \alpha x + C,$$

where C is a constant. Now taking the exponential of both sides gives

$$|y| = e^{\alpha x + C} = e^{\alpha x}e^{C} = Ae^{\alpha x},$$

where $A = e^{C}$ is a positive constant. If $y > 0$, $y = Ae^{\alpha x}$, and if $y < 0$, $y = -Ae^{\alpha x}$, so we can drop the modulus sign and write $y = Ae^{\alpha x}$, where now the constant A can be positive, negative or zero.

The example (2.13) is the most important differential equation in this book! It comes up many times in the following chapters. Rather than going through the process of solving this equation each time it comes up, it is best just to learn that the general solution of $y' = \alpha y$ is $y = Ae^{\alpha x}$.

---

[2] The step from (2.11) to (2.12) is not mathematically rigorous, but it can be justified by introducing a couple more steps and using a change of variable.

Example 2.8 Find the solution of

$$\dot{x} = -x^2 t^2$$

with the initial condition $x(0) = 3$.

This equation is separable. As in the previous example, notice that $x = 0$ is a solution. For $x \neq 0$, applying the separation of variables method gives

$$\int -\frac{1}{x^2}\, dx = \int t^2 \, dt.$$

These are both easy integrals, and the result is

$$\frac{1}{x} = \frac{t^3}{3} + C \implies x = \frac{3}{t^3 + 3C},$$

where C is a constant. To find the specific solution that obeys the initial condition, we set $x = 3$ at $t = 0$, giving $C = 1/3$. The specific solution is therefore

$$x = \frac{3}{t^3 + 1}.$$

As mentioned in Remark 2.1, it is wise to check that our solution obeys the original differential equation and initial condition. For the specific solution found,

$$\dot{x} = \frac{-3}{(t^3 + 1)^2} \times 3t^2 = -\frac{9t^2}{(t^3 + 1)^2} = -x^2 t^2,$$

so the differential equation is satisfied. Also, at $t = 0$, $x = 3$, so the solution obeys the initial condition.

2.3 First-Order Linear Equations

Another case where there is a systematic process for finding the solution occurs when the differential equation is first-order and linear. Any first-order linear equation can be written in the form

$$\frac{dy}{dx} + p(x)y = q(x), \tag{2.14}$$

for some functions $p(x)$ and $q(x)$.

Before setting out the general method, it is helpful to consider a simple example that illustrates the key idea of the method.

Example 2.9 Solve the differential equation

$$x\frac{dy}{dx} + y = \cos x. \tag{2.15}$$

This first-order equation is linear, as there are no products or nonlinear functions of the dependent variable y. It could be written in the form (2.14) by first dividing through by x, so $p(x) = 1/x$, $q(x) = \cos x/x$. It is not separable, since dy/dx cannot be written as the product of a function of x and a function of y.

We can solve (2.15) by a simple trick. Consider the function $f(x) = xy(x)$. Differentiating this, using the product rule, gives

$$\frac{df}{dx} = x\frac{dy}{dx} + y,$$

which is exactly the left-hand-side (LHS) of (2.15). So (2.15) can be written as

$$\frac{d(xy)}{dx} = \cos x,$$

which can be solved just by integrating both sides to get

$$xy = \sin x + C.$$

Hence the general solution is

$$y = \frac{\sin x + C}{x},$$

where C is an arbitrary constant.

The trick used in Example 2.9 cannot be applied directly to the general first-order linear differential equation (2.14). But it can be used if we first multiply through by an appropriate factor, called the *integrating factor*. The integrating factor for (2.14) is $I(x) = \exp(\int p(x)dx)$. Using the properties of the exponential function and the chain rule, it follows that the derivative of $I(x)$ is

$$\frac{dI(x)}{dx} = \exp\left(\int p(x)dx\right) \times \frac{d(\int p(x)dx)}{dx} = I(x)p(x).$$

Multiplying (2.14) through by $I(x)$ gives

$$I(x)\frac{dy}{dx} + I(x)p(x)y = I(x)q(x).$$

Now the LHS can be written as the derivative of a product as in Example 2.9.

$$\frac{d(I(x)y)}{dx} = I(x)q(x).$$

This follows from the product rule and the derivative of $I(x)$ calculated above. Now we can integrate both sides of the equation and we have the general solution, if we can do the integral. There will, as usual, be a constant of integration in the solution.

Example 2.10 Find the general solution of

$$x\frac{dy}{dx} + 4y = x^2. \tag{2.16}$$

This equation is first-order and linear, but it is not in the standard form (2.14) because the coefficient of the derivative is not 1. To get the standard form, we first need to divide through by x:

$$\frac{dy}{dx} + \frac{4}{x}y = x. \tag{2.17}$$

This is in the form (2.14), with $p(x) = 4/x$ and $q(x) = x$. The integrating factor is

$$\exp\left(\int \frac{4}{x}\,dx\right) = \exp(4\log x) = \exp(\log(x^4)) = x^4.$$

We do not need to include a constant of integration when finding the integrating factor. If we did include it, the effect would only be to multiply the equation through by a constant. Multiplying (2.17) by the integrating factor x^4 gives

$$x^4\frac{dy}{dx} + 4x^3y = x^5. \tag{2.18}$$

According to the general case above, the LHS here should be the derivative of x^4y, the integrating factor multiplied by the unknown function $y(x)$. It's easy to check that this is the case. So (2.18) can be written as

$$\frac{d(x^4y)}{dx} = x^5.$$

The last step is to integrate both sides, including a constant of integration:

$$x^4y = \frac{x^6}{6} + C \;\Rightarrow\; y = \frac{x^2}{6} + \frac{C}{x^4}.$$

2.4 Second-Order, Linear, Constant-Coefficient Equations

A third type of differential equation where there is a systematic procedure for finding an analytical solution is when the equation is second-order and linear and the coefficients of the dependent variable are all constants.

Consider first the equation

$$a\frac{d^2y}{dx^2} + b\frac{dy}{dx} + cy = 0, \tag{2.19}$$

where a, b and c are constants. This equation is linear and *homogeneous*, as defined in Sect. 2.1.4, since every term includes the dependent variable y or one of its derivatives once. The method of solving (2.19) is motivated by Example 2.13. In that case the equation was first-order, linear and homogeneous with constant coefficients, and the solution was an exponential function. So for (2.19), the method is to guess that the solution will be an exponential function. We try a solution of the form $y = Ae^{mx}$, where A and m are constants. Substituting this guess into (2.19) gives

$$aAm^2e^{mx} + bAme^{mx} + cAe^{mx} = 0.$$

We can cancel the common factors of A and e^{mx}, but first we should check if these quantities can be zero. A can be zero, which gives the valid but not very interesting solution $y = 0$. The exponential function can never be zero. So, for a non-zero solution, the constant m must obey the *auxiliary equation*,

$$am^2 + bm + c = 0. \tag{2.20}$$

There are then three possibilities, depending on the roots of (2.20). These three cases are considered in the following three sections.

2.4.1 Real, Distinct Roots

Suppose that (2.20) has two real roots, m_1 and $m_2 \neq m_1$. Then $y = e^{m_1 x}$ and $y = e^{m_2 x}$ are both solutions of (2.19). By the superposition principle (Sect. 2.1.4), the general solution of (2.19) in this case is

$$y = Ae^{m_1 x} + Be^{m_2 x}. \tag{2.21}$$

This solution contains two arbitrary constants, A and B, as expected for a second-order differential equation.

2.4.2 Equal Roots

Now consider the case when (2.20) has two equal roots, in other words, a double root, $m = m_1$. Then we have a solution of the equation, $y = Ae^{m_1 x}$, but this solution only contains one arbitrary constant. In general there should be two constants, or equivalently, two independent solutions of the equation. In this case, it can be shown (see Exercise 2.5) that the second solution is $xe^{m_1 x}$, so the general solution is

$$y = Ae^{m_1 x} + Bxe^{m_1 x}. \tag{2.22}$$

2.4.3 Complex Roots

The third possibility is that the roots of (2.20) may be complex. If this is the case, they must form a complex conjugate pair,

$$m_1 = m_r + im_i, \quad m_2 = m_r - im_i,$$

where $i^2 = -1$. The solution (2.21) is mathematically valid, but in general it is complex. In most applications, we are looking for a real solution. Using the complex conjugate forms of m_1 and m_2, (2.21) is

$$y = Ae^{m_r x}e^{im_i x} + Be^{m_r x}ee^{-im_i x}.$$

This solution is real if A and B are complex conjugates of each other. Using Euler's formula $e^{i\theta} = \cos\theta + i\sin\theta$ the solution can be written as

$$y = (A + B)e^{m_r x}\cos(m_i x) + (A - B)e^{m_r x}i\sin(m_i x)$$
$$= 2A_r e^{m_r x}\cos(m_i x) - 2A_i e^{m_r x}\sin(m_i x),$$

where A_r and A_i are the real and imaginary parts of A. With a simple rescaling of the constants, the real solution of (2.19) in the case when the auxiliary equation has complex roots is

$$y = Ce^{m_r x}\cos(m_i x) + De^{m_r x}\sin(m_i x)$$
$$= e^{m_r x}\big(C\cos(m_i x) + D\sin(m_i x)\big),$$

where C and D are arbitrary constants. The solution oscillates, and its amplitude either decreases or increases exponentially, or remains constant, depending on the value of m_r.

Remark 2.2 The method described in this section can be applied to a linear, constant-coefficient, homogeneous differential equation of any order. For example, if the equation is third-order, the auxiliary equation is a cubic, and there are generally three possible solutions, leading to a general solution with three constants.

2.4.4 The Inhomogeneous Case

Now consider an *inhomogeneous* second-order, linear, constant-coefficient differential equation. The general form of such an equation is

$$a\frac{d^2y}{dx^2} + b\frac{dy}{dx} + cy = f(x), \qquad (2.23)$$

where $f(x)$ may be any function of x, but a, b and c are still constant. Because the equation is not homogeneous, the superposition principle no longer holds: in general, the sum of two solutions of (2.23) is not a solution of the same equation.

The method uses three steps. The first step is to solve the homogeneous problem in which $f(x) = 0$, using the method of Sect. 2.4. The solution found is called the *complementary function*. It will contain two arbitrary constants.

The second step is to find what is called the *particular integral*. This is a function that satisfies (2.23). It is obtained by a process of trial and error, that gets easier with practice. First, make a guess at the solution. The guess is inspired by the form of $f(x)$. If $f(x)$ is an exponential function, for example e^{3x}, try a function of the same form, $y = Ce^{3x}$, substitute into (2.23) and see if there is a value of C for which the equation is satisfied. If $f(x)$ is a polynomial function, try a polynomial of the same degree. For example, if $f(x) = x^2$, try $y = C_2x^2 + C_1x + C_0$, substitute in and try to solve for the constants. If $f(x)$ is a trig function, try a combination of trig functions, so if $f(x) = \sin x$, try $y = C_1 \cos x + C_2 \sin x$. This part of the solution, the particular integral, should not have any arbitrary constants—they should all be determined by substituting into the differential equation and comparing similar terms.

The final step is to add together these two parts. The general solution of (2.23) is the complementary function plus the particular integral. This method is best illustrated through examples.

Example 2.11 Find the general solution of

$$\frac{d^2y}{dx^2} + 3\frac{dy}{dx} - 4y = f(x), \qquad (2.24)$$

In the cases

(a) $f(x) = 0$,
(b) $f(x) = e^{2x}$,
(c) $f(x) = 4x^2$.

(a) In the homogeneous case, $f(x) = 0$, trying a solution $y = Ae^{mx}$ gives the auxiliary equation for m,

$$m^2 + 3m - 4 = 0 \Rightarrow (m + 4)(m - 1) = 0.$$

The solutions are $m = -4$ and $m = 1$, so the general solution is

$$y = Ae^{-4x} + Be^x,$$

where A and B are arbitrary constants.

(b) In this case, first find the complementary function, which is the function found in part (a). Now try to find the particular integral. Since $f(x) = e^{2x}$, try a function $y = Ce^{2x}$. Substituting this into (2.24) gives

$$4Ce^{2x} + 6Ce^{2x} - 4Ce^{2x} = e^{2x},$$

so the solution is $C = 1/6$. The general solution is then the complementary function plus the particular integral,

$$y = Ae^{-4x} + Be^x + \frac{1}{6}e^{2x}.$$

(c) For the case $f(x) = 4x^2$, the complementary function is the same as in (b). For the particular integral, try $y = C_2x^2 + C_1x + C_0$. Even though $f(x)$ only has a quadratic term, all three terms are needed here, because differentiation introduces lower powers of x. Substituting this into (2.24) gives

$$2C_2 + 3(2C_2x + C_1) - 4(C_2x^2 + C_1x + C_0) = 4x^2.$$

Equating coefficients of x^2 gives $C_2 = -1$. Then the coefficients of x give $6C_2 - 4C_1 = 0$, so $C_1 = -3/2$, and the constant terms give $2C_2 + 3C_1 - 4C_0 = 0$, so $C_0 = -13/8$. Hence the general solution is

$$y = Ae^{-4x} + Be^x - x^2 - \frac{3}{2}x - \frac{13}{8}.$$

Remark 2.3 In addition to the main types of particular integral discussed above, there are a few special cases. One special case arises when the inhomogeneous term $f(x)$ is a multiple of one of the terms in the complementary function. For example, consider Example 2.11 in the case $f(x) = e^x$. In this case, trying a particular integral $y = Ce^x$, similar to the method for part (b), does not work; there is no solution for C. The trick to get around this difficulty is similar to that for the case of repeated roots in a complementary function calculation: include an extra factor of x. When $f(x) = e^x$, trying $y = Cxe^x$ in (2.24) works, giving the answer $C = 1/5$.

2.5 Coupled Linear Systems

In many applications of differential equations, there is more than one dependent variable, each with its own differential equation, and the equations are coupled together, so the rate of change of one variable depends on the value of another. Examples include models of competition between species and models for the spread of an infection, both of which will be investigated in Chap. 5.

Two linear, constant-coefficient, coupled equations can be solved by an elimination method, leading to a single second-order equation that can then be solved using the methods of Sect. 2.4. This method is illustrated by Example 2.12 below. A more elegant way of solving coupled linear systems will be introduced in Chap. 4.

Example 2.12 Solve the system

$$\dot{x} = x - 2y, \tag{2.25}$$

$$\dot{y} = x + 4y. \tag{2.26}$$

In these equations, it is implicit that x and y are both dependent variables, functions of an independent variable that we may call t. To solve this, first differentiate either one of the equations with respect to t. Differentiating (2.25) gives

$$\ddot{x} = \dot{x} - 2\dot{y}.$$

We can eliminate $\dot{y}$ using (2.26), to find

$$\ddot{x} = \dot{x} - 2x - 8y.$$

Now y still appears in the equation, but this can be eliminated using (2.25) to write

$$y = \frac{x - \dot{x}}{2}. \tag{2.27}$$

This leads to a second-order differential equation for x,

$$\ddot{x} = \dot{x} - 2x - 4(x - \dot{x}) = 5\dot{x} - 6x.$$

This can now be solved as described in Sect. 2.4, giving the result

$$x = Ae^{2t} + Be^{3t},$$

where A and B are constants. Finally, $y(t)$ can be found directly from (2.27), in terms of the same two constants A and B,

$$y = -\frac{A}{2}e^{2t} - Be^{3t}. \tag{2.28}$$

The process here is similar to solving two simultaneous equations, but the difference is that two substitutions are needed, one to eliminate $\dot{y}$ and another to eliminate y.

2.6   Other Methods and Tricks

The methods outlined in this chapter are the main techniques used for solving differential equations analytically. There are no general methods for obtaining analytical solutions of more complicated differential equations, especially when they are nonlinear. So an alternative approach is needed. One such approach is the qualitative method, which is the subject of the next chapter and most of the rest of this book.

However, there are several tricks that work for particular types of equation, either providing a direct solution, or using a substitution to turn the equation into a type that we can solve. The following examples show two such methods.

Example 2.13 Show that the substitution $y = 1/u$ transforms the nonlinear equation

$$y' + p(x)y = q(x)y^2 \tag{2.29}$$

into a linear equation for $u(x)$. Hence solve $y' + y = e^x y^2$.

Set $y = 1/u$, so that $y' = (-1/u^2)u'$. Then (2.29) becomes

$$-\frac{u'}{u^2} + \frac{p(x)}{u} = \frac{q(x)}{u^2} \;\Rightarrow\; -u' + p(x)u = q(x).$$

This equation is in the standard first-order linear form (2.14) so it can be solved using the methods of Sect. 2.3. For $y' + y = e^x y^2$, the change of variable $y = 1/u$ gives $-u' + u = e^x$. This is solved using the integrating factor e^{-x} to give $u = (C - x)e^x$, where C is a constant, so the general solution of (2.29) is

$$y = \frac{e^{-x}}{C - x}.$$

This method can be generalised to the case where the right-hand side includes a term y^n, see Exercise 2.11.

Example 2.14 By seeking a solution of the form $y = Ax^p$, solve

$$x^2 y'' - 5xy' + 5y = 0. \tag{2.30}$$

Trying the suggested form for y gives

$$x^2 Ap(p - 1)x^{p-2} - 5xApx^{p-1} + 5Ax^p = 0.$$

Every term has a factor of Ax^p, and cancelling this factor (after noting that $A = 0$ gives a solution $y = 0$) gives a quadratic equation, $p^2 - 6p + 5 = 0$. Hence $p = 1$ or $p = 5$. Since (2.30) is linear and homogeneous, the superposition principle holds and the general solution is

$$y = Ax + Bx^5.$$

This method works for linear homogeneous equations if the nth derivative is multiplied by x^n in each term. Note that this method is similar to the method of Sect. 2.4.

Key Points from Chap. 2

- A differential equation is a relationship between a function and its derivatives.
- The highest derivative appearing in a differential equation gives the *order* of the equation.
- A differential equation is *linear* if it only includes linear functions of the dependent variable, and contains no products of these terms. Otherwise, it is *nonlinear*.
- The general solution of an nth-order differential equation includes n arbitrary constants.
- These constants can be found if we have the appropriate number of *initial conditions* or *boundary conditions*.
- A first-order differential equation is *separable* if it can be written in the form $dy/dx = f(x)g(y)$.
- A separable differential equation can be solved by separating the variables to each side of the equation and then integrating.
- A first-order, linear differential equation $dy/dx + p(x)y = q(x)$ can be solved by multiplying through by the *integrating factor*, $I(x) = \exp(\int p(x)\, dx)$.
- A second-order, linear, *homogeneous* differential equation with constant coefficients, $ay'' + by' + cy = 0$, can be solved by seeking a solution $y = Ae^{mx}$ and solving the resulting quadratic equation for m.
- A second-order, linear, *inhomogeneous* differential equation with constant coefficients, $ay'' + by' + cy = f(x)$, is solved by first solving the homogeneous equation, and then adding the *particular integral*, found by trying a guess closely related to $f(x)$.
- Two coupled first-order differential equations can be solved by eliminating one of the dependent variables, writing the system of two equations as a single second-order equation.

Exercises

2.1 For each of the following differential equations, write down the order of the equation and state whether it is linear or nonlinear. If the equation is linear, is it homogeneous or inhomogeneous?

(a)

$$\frac{d^2 y}{dx^2} + x^2 \frac{dy}{dx} = \sin x,$$

(b)

$$\dot{x} + x^2 t = 1 + t,$$

(c)

$$y''' + y'' + yy' = x,$$

(d)

$$\ddot{u} + 4\dot{u} + u = 0.$$

2.2 Which of the following first-order differential equations are separable, and which are linear? Apply a suitable method to find the general solution of each.

(a)

$$\frac{dy}{dx} = y + y^2,$$

(b)

$$\frac{dy}{dx} + y = e^x,$$

(c)

$$\frac{dy}{dx} + 2xy = x.$$

2.3 Find the function $x(t)$ that obeys the differential equation

$$x\dot{x} = t(x^2 + 1)$$

and the initial condition $x(0) = 1$. How does the solution change if the initial condition is $x(0) = -1$?

2.4 Find the general solution of

$$x\frac{dy}{dx} - 3y = x^3,$$

and the specific solution that obeys the boundary condition $y(1) = 2$.

2.5 Verify that for the second-order differential equation (2.19), the solution (2.22) is correct in the case where the auxiliary equation (2.20) has equal roots.

2.6 Find the general solution of the differential equation

$$\frac{d^2y}{dx^2} + b\frac{dy}{dx} + y = 0$$

in the cases

(a) $b = 0$,
(b) $b = 2$,
(c) $b = 4$.

2.7 Find the general solution of the fourth-order differential equation

$$\frac{d^4y}{dx^4} - 13\frac{d^2y}{dx^2} + 36y = 0.$$

Hint: see Remark 2.2.

2.8 Find the general solution of

$$\frac{d^2y}{dx^2} + 4y = f(x)$$

in the cases

(a) $f(x) = \cos x$,
(b) $f(x) = \cos 2x$,
(c) $f(x) = x$,
(d) $f(x) = x^3$.

2.9 Repeat the problem of Example 2.12, but eliminate x to find a second-order equation for $y(t)$. Also find the specific solution that obeys the initial conditions $x(0) = 2$, $y(0) = 5$.

2.10 Find the general solution of the system

$$\dot{x} = x - 2y,$$
$$\dot{y} = 5x + 3y.$$

2.11 Show that the substitution $u = y^{1-n}$ transforms the nonlinear equation

$$y' + p(x)y = q(x)y^n$$

into a linear equation for $u(x)$. Hence find the general solution of

$$xy' + y = \frac{2x^3}{y^2}.$$

2.12 (a) Find the general solution of

$$x^2 y'' - 2y = 0.$$

(b) Find the specific solution that is finite at $x = 0$ and obeys the condition $y(1) = 1$.

Chapter 3
Qualitative Methods for First-Order Differential Equations

This chapter introduces the idea of the qualitative approach to differential equations. This means that rather than find an explicit solution, which is often not possible, the aim is to describe the behaviour of solutions to the equation, and sketch the possible solutions for different initial conditions. This approach will be the main theme of this book. For first-order differential equations, the qualitative method involves the construction of a phase line, which gives a visual representation of the behaviour of solutions of the equation.

3.1 Introduction to Qualitative Methods

The previous chapter summarises the type of material that is usually included in introductory undergraduate courses on differential equations. But this can create a misleading impression. It may appear that there is a sequence of methods for solving different types of differential equations, and that more complicated differential equations can be solved by using more advanced methods. But that is not the case. In fact, many differential equations of interest in applications, particularly nonlinear ones, cannot be solved analytically. There is no way to just write down a function that obeys the differential equation.

When faced with a differential equation that cannot be solved analytically, there are several alternative approaches. One important method is to approximate the differential equation by a discrete system, or difference equation, and then solve this numerically using a computer. Chapter 7 investigates the behaviour of difference equations. This method has been very successful in many contexts. For example, it is used in weather forecasting, where accurate results can be obtained over short time scales, but long-term forecasts are unreliable because of the phenomenon of chaos, as discussed in Chap. 8.

© The Author(s), under exclusive license to Springer Nature Switzerland AG 2025
P. C. Matthews, *Differential Equations, Bifurcations and Chaos*,
Springer Undergraduate Mathematics Series,
https://doi.org/10.1007/978-3-031-99543-9_3

Another approach may be useful if the differential equation has a small parameter in it. If we can solve the equation when the parameter is set to zero, it may be possible to find a good approximation to the solution when the parameter is small but not zero. This is known as the perturbation method, or the method of asymptotic expansion.

But the main aim of this book is to introduce the concept of a *qualitative* approach. The idea here is to provide a description of the solution to the differential equation, without actually solving it. For example, suppose that we have a differential equation for a function $x(t)$, and an initial condition $x(0)$. We may be able to say, simply by looking at the differential equation, that $x(t)$ increases to a maximum, then decreases, and that $x(t)$ tends to zero as t tends to infinity. Hence we may be able to describe the solution $x(t)$, and sketch a graph of it, even though we cannot write down an explicit formula for it. This is what is meant by the term *qualitative*—we can describe the qualities of the solution.

In contrast, if we can find an exact analytical solution, for example, if we know that $x = 2.75$ when $t = 1$, that is known as a *quantitative* solution.

The following simple example illustrates the idea of the qualitative method.

Example 3.1 Give a qualitative description of the solutions to the differential equation

$$\frac{dx}{dt} = 1 - x^2. \tag{3.1}$$

This is a first-order differential equation for $x(t)$, and it is *autonomous*—the independent variable t does not appear explicitly in the equation. It is also separable, and it can be solved using the methods of the previous chapter, see Sect. 2.2. However, we can describe and sketch solutions without having to solve the equation. Of course, an initial condition $x(0)$ is needed in order to find a specific solution, otherwise the solution contains an arbitrary constant.

The equation tells us that the slope of the function $x(t)$ is equal to $1 - x^2$. Therefore, if the initial condition $x(0)$ lies between -1 and 1, $\dot{x}(0) > 0$, so x is initially an increasing function of t. But if $x(0) > 1$ or $x(0) < -1$, then $\dot{x}(0) < 0$ and x starts to decrease. The borderline cases are particularly important. Suppose that $x(0) = 1$. Then $\dot{x}(0) = 0$, so the graph of $x(t)$ has slope zero and the function $x(t)$ does not change. Therefore, $x(t)$ remains at the value 1. Alternatively, we can see that $x(t) = 1$ for all t is a solution of the equation. A solution of this type is called a *fixed point* or *equilibrium point* of the differential equation. Equation (3.1) has two fixed points, $x = 1$ and $x = -1$.

The two fixed points of (3.1) are different in character. Consider what happens to solutions to the equation near one of the fixed points. Suppose that at some time t, x is just greater than 1. In that case, the equation tells us that $\dot{x} < 0$, so x will decrease, moving closer to the fixed point $x = 1$. If x is just less than 1, $\dot{x} > 0$ so x will increase and will again move closer to the point $x = 1$. A fixed point of this type, where solutions that start near the fixed point move towards it, is said to be

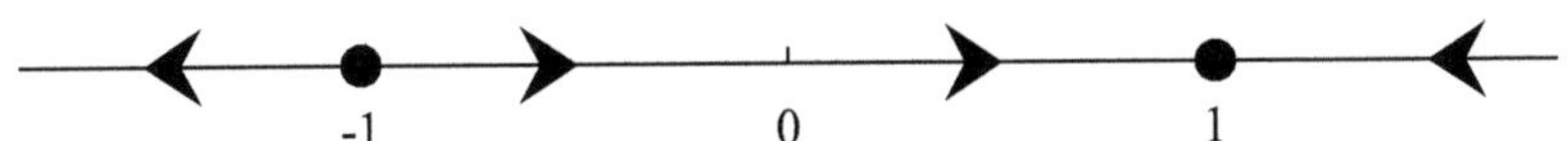

Fig. 3.1 Phase line for the differential equation (3.1). The large dots are the fixed points, $x = -1$ and $x = 1$. Arrows indicate that x increases in the interval $-1 < x < 1$, but decreases outside this range

stable. Now suppose that x is slightly greater than -1. Then (3.1) shows that $\dot{x} > 0$, so x will increase, moving away from the point $x = -1$. If x is just less than -1, we have $\dot{x} < 0$ so x decreases and again x moves away from the point $x = -1$. In this case, the fixed point is said to be *unstable*, since solutions move away from it.

Using this information, we can describe the solutions of (3.1) for all initial conditions $x(0)$. If $x(0) = 1$, then $x(t) = 1$ for all t, and similarly, if $x(0) = -1$, then $x(t) = -1$ for all t. If $-1 < x(0) < 1$, so the initial condition lies between the two fixed points, then $\dot{x} > 0$ and $x(t)$ increases and approaches the fixed point at $x = 1$, so $x(t) \to 1$ as $t \to \infty$. If $x(0) > 1$, $\dot{x} < 0$ and x decreases, again approaching $x = 1$ as t approaches infinity. Finally, if $x(0) < -1$, $\dot{x} < 0$, so x decreases, making $\dot{x}$ more negative, and in that case $x(t) \to -\infty$.

This qualitative description can be summarised in the *phase line* for the differential equation, which is shown in Fig. 3.1. The line represents the variable $x(t)$, and the arrows show how x changes as t increases, so arrows point to the right in the region $-1 < x < 1$ where $\dot{x} > 0$, and to the left outside this region. Note that the arrows point away from the unstable fixed point at $x = -1$, and towards the stable fixed point at $x = 1$. Using the phase line we can also sketch the solutions $x(t)$, see Example 3.6 and Fig. 3.4.

3.2   Fixed Points and Their Stability

This section defines more generally and precisely the terms used in Example 3.1.

Definition 3.1 Consider a first-order, autonomous differential equation,

$$\frac{dx}{dt} = f(x). \tag{3.2}$$

A point x_0 is a *fixed point* of (3.2) if and only if $f(x_0) = 0$.

Fixed points are also known as *equilibrium points* or *critical points*. A fixed point x_0 is a solution of (3.2) in which x remains constant. If $x(t) = x_0$, both sides of (3.2) are zero.

An important constraint on the solutions of a first-order, autonomous differential equation is that no solution can pass through a fixed point, since if it did so, $\dot{x}$ would have to be non-zero at the fixed point, which is a contradiction. Therefore, all non-constant solutions must be confined to values of x that are either between two fixed points, or greater than the largest fixed point, or less than the lowest fixed point.

3.2.1 Stable and Unstable Fixed Points

The concept of the stability of fixed points is very important in mathematics and its applications.

When a system is perturbed slightly from its equilibrium state, the fixed point is *stable* if the perturbation decreases, so that the system returns to the fixed point, and it is *unstable* if the perturbation grows, leading to solutions that move away from the fixed point. In any real application, small perturbations will always be present, so stable fixed points are the ones that are observed, while unstable ones are either not seen at all, or seen only for a short time.

Example 3.2 A simple physical example illustrating stable and unstable fixed points is a pendulum suspended from a pivot point. Let x be the angle the pendulum rod makes with the downward vertical. This system has two fixed points. One is $x = 0$, where the pendulum is hanging vertically below its pivot point. This is a stable fixed point, since if the pendulum is disturbed, it will oscillate and then, due to friction and air resistance, return to rest in its original position. But there is a second fixed point, $x = \pi$, where the pendulum is pointing vertically upwards. This second fixed point is unstable. If we allow x to take all real values, there are fixed points at $x = n\pi$, for any integer n, but there are only two physically distinct fixed points. This system, which is modelled mathematically by a nonlinear, second-order differential equation, will be studied in Chap. 5. A simpler first-order model is given in Example 3.3 below.

To investigate the stability of a fixed point of a first-order differential equation mathematically, suppose that (3.2) has a fixed point at $x = x_0$, so $f(x_0) = 0$. Now suppose that x is close to x_0. Define a new variable $y(t)$ to represent the perturbation of x from x_0, so

$$y(t) = x(t) - x_0. \tag{3.3}$$

The differential equation for $y(t)$ is

$$\frac{dy}{dt} = \frac{dx}{dt} = f(x) = f(x_0 + y). \tag{3.4}$$

Now assume that the function f has a Taylor series expansion, so we can write this equation as

$$\frac{dy}{dt} = f(x_0) + yf'(x_0) + \frac{y^2}{2}f''(x_0) + O(y^3), \tag{3.5}$$

where $O(y^3)$ represents terms proportional to y^3 or smaller, see Appendix A.1. The first term in the expansion is zero, since x_0 is a fixed point.

Since we are considering a small perturbation to the fixed point x_0, y is a small quantity. Hence the y^2 term in (3.5) is in general much smaller than the y term. If y is very small, a good approximation is obtained by neglecting the y^2 and higher terms. This approximation is known as *linearisation*. Making this approximation leads to

$$\frac{dy}{dt} = f'(x_0)y. \tag{3.6}$$

This is known as the *linearised* form of (3.5), since the equation is now linear. The factor $f'(x_0)$ that appears in the equation is just a constant, depending on the function f and the value of x_0.

The solution to (3.6) is simply an exponential function, see Example 2.7,

$$y(t) = y_0 e^{f'(x_0)t}, \tag{3.7}$$

where y_0 is the initial condition for y. This solution shows that if $f'(x_0) > 0$, then $y(t)$ grows exponentially, meaning that $x(t)$ moves away from x_0 exponentially, so the fixed point is unstable. But if $f'(x_0) < 0$, the perturbation decays exponentially and we have a stable fixed point.[1]

Definition 3.2 A fixed point x_0 of the autonomous differential equation (3.2) is *stable* if $f'(x_0) < 0$. It is *unstable* if $f'(x_0) > 0$.

Returning to Example 3.1, in that case, $f(x) = 1 - x^2$, so $f'(x) = -2x$. For the fixed point $x_0 = -1$, $f'(-1) = 2 > 0$, so this point is unstable, but the fixed point $x_0 = 1$ is stable, because $f'(1) = -2 < 0$. This is in agreement with the result obtained by considering the sign of $f(x)$ in Example 3.1, and with the phase line diagram in Fig. 3.1.

Example 3.3 Find all the fixed points of the differential equation

$$\frac{dx}{dt} = -\sin x \tag{3.8}$$

and determine their stability.

The fixed points are points where $-\sin x = 0$, so $x = n\pi$, where n is any integer. To find the stability of the fixed points, set $f(x) = -\sin x$ and find $f'(x) = -\cos x$ at each of the fixed points. When $x = n\pi$, $f'(x) = -\cos n\pi = (-1)^{n+1}$, so the fixed points with n even ($x = 0, \pm 2\pi, \pm 4\pi \ldots$) have $f'(x) = -1$ and are stable, while the fixed points with n odd ($x = \pm\pi, \pm 3\pi, \ldots$) are unstable, with $f'(x) = +1$. This example can be thought of as a simplified case of the pendulum

---

[1] The fact that (3.6) arises from consideration of the stability of a fixed point is the reason why (2.13) was described as the most important differential equation in this book.

problem discussed in Example 3.2 above, in the case where the damping of the pendulum is very strong.

Example 3.4 The *logistic model* of population growth is

$$\frac{dx}{dt} = rx - bx^2, \tag{3.9}$$

where r and b are positive constants. Show that by rescaling the variables x and t, we can set $r = b = 1$. Find the fixed points of the model, find whether they are stable or unstable, and discuss the interpretation of the terms in the equation and the fixed points in terms of a biological population.

Suppose that we rescale $x(t) = \alpha X(t)$ and $t = \beta T$. Then the logistic equation becomes

$$\frac{\alpha}{\beta}\frac{dX}{dT} = r\alpha X - b\alpha^2 X^2. \tag{3.10}$$

Now if we choose α and β so that $\alpha/\beta = r\alpha = b\alpha^2$, all the constants will drop out of the equation. This can be achieved by setting $\beta = 1/r$ and $\alpha = r/b$. This process is called *nondimensionalisation*, since the new variables X and T do not have dimensions. It is a useful technique for reducing the number of parameters in a problem. The nondimensionalised equation is then

$$\frac{dX}{dT} = X - X^2 = X(1 - X) = f(X), \tag{3.11}$$

so the fixed points are $X = 0$ and $X = 1$. Since $f'(X) = 1 - 2X$, $f'(0) = 1$ and $f'(1) = -1$, so $X = 0$ is an unstable fixed point, while $X = 1$ is stable.

The interpretation of the model in terms of a population of, for example, biological cells or animals, is as follows. The fixed point $X = 0$ represents extinction of the population. If there are no animals at time $T = 0$, this will remain the case for all T. For any population model representing a closed system, $X = 0$ should be a fixed point. The point $X = 1$ represents a stable population level. If the population X is very small, the linear term dominates, since X^2 is much less than X, and the population grows exponentially, until it becomes large enough for the nonlinear term $- X^2$ to become significant. This term prevents the growth continuing to infinity, which would be unrealistic. The quadratic term could represent competition between animals, for example for food, which limits the population growth. For any positive initial condition $X(0)$, the population approaches $X = 1$ as $T \to \infty$ Since (3.11) is intended to be a population model, only solutions with $X > 0$ are of interest.

The logistic equation is widely used to model biological systems, and there are many possible modifications and extensions of it. For example, the competition between two different species can be modelled by coupling two logistic equations together (see Chap. 5).

3.2.2 Special Cases

The previous section has shown that a fixed point x_0 is stable if $f'(x_0) < 0$ and unstable if $f'(x_0) > 0$. But what happens in the borderline case, when $f'(x_0) = 0$? In that case, the linearised equation (3.6) seems to show that the perturbation y does not change. But this is incorrect in general. The assumption made after (3.5), that the y term is much greater than the y^2 term, is not valid in the case $f'(x_0) = 0$, so we need to consider further terms in the Taylor expansion (3.5).

If $f(x_0) = f'(x_0) = 0$ but $f''(x_0) \neq 0$, then the dominant term in (3.5) for small y is the quadratic term, so the equation may be approximated by

$$\frac{dy}{dt} = \frac{y^2}{2} f''(x_0). \tag{3.12}$$

Suppose that $f''(x_0) > 0$. Then for all $y \neq 0$, the right-hand side of (3.12) is positive, so y increases as a function of t. Hence if $y < 0$, y increases towards $y = 0$, showing that the fixed point is stable to this type of perturbation. But if $y > 0$, y increases away from $y = 0$, showing unstable behaviour. If $f''(x_0) < 0$, the picture is reversed, so any non-zero y decreases, and solutions with $y > 0$ decay while solutions with $y < 0$ grow in magnitude. In either case, some perturbations increase, and others decrease. A fixed point of this type, where some perturbations grow and some decay, is regarded as unstable. This makes sense physically, because in most real situations, small perturbations in different directions are likely to be present.

If a fixed point has $f'(x_0) = f''(x_0) = 0$, but $f'''(x_0) \neq 0$, then the behaviour of small perturbations is governed by the cubic term in the Taylor expansion,

$$\frac{dy}{dt} = \frac{y^3}{6} f'''(x_0). \tag{3.13}$$

The situation here is more similar to the linearised case. If $f'''(x_0) < 0$ then the right-hand side of (3.13) is negative when y is positive, and positive when y is negative, so perturbations decrease and solutions approach $y = 0$, so the fixed point is stable. A significant difference from the linear case is that the approach to $y = 0$ is not exponential, but proportional to a negative power of t, see Exercise 3.5. If $y'''(x_0) > 0$ then perturbations grow and the fixed point is unstable.

In the general case, if the first non-zero derivative at a fixed point is the nth derivative $f^{(n)}(x_0)$ then the appropriate approximation is

$$\frac{dy}{dt} = \frac{y^n}{n!} f^{(n)}(x_0). \tag{3.14}$$

If n is even the fixed point is always unstable, like the quadratic case (3.12), but if n is odd the fixed point is stable or unstable depending on the sign of the derivative, as in the cubic case (3.13).

3.3   Phase Lines and Solution Curves

The main aim of the qualitative approach to solving differential equations is to produce diagrams illustrating the behaviour of the solutions of the equation. For an autonomous, first-order differential equation, this behaviour can be summarised in a one-dimensional diagram, the phase line. Using the phase line, we can sketch graphs of the solution curves to the differential equation.

3.3.1   Phase Lines

The idea of the phase line for the autonomous differential equation (3.2) was introduced in Example 3.1. The phase line represents the dependent variable, x in (3.2). The independent variable t is not plotted, but the change of x with t is shown by arrows on the phase line. One way to interpret the phase line is to imagine that the equation represents the motion of a particle along the x axis.

To construct the phase line, first find all of the fixed points of the equation, in other words, find the values of x for which $f(x) = 0$ in (3.2). In some cases, it may not be possible to find these points analytically, but if we can determine how many fixed points there are, and their approximate location, it is still possible to plot a qualitative phase line. Fixed points are marked by dots on the phase line.

The second step in sketching the phase line is to find the sign of $\dot{x}$ in each region of the phase line. Assuming that the function f is continuous, $\dot{x}$ can only change its sign at a fixed point, so we only need to check the sign of $\dot{x}$ at one location between any two fixed points. In the usual case, where $f'(x_0) \neq 0$ at a fixed point, the sign of $\dot{x}$ changes as we pass through a fixed point, but this may not be true in the special cases described in Sect. 3.2.2.

The sign of $\dot{x}$ is indicated by marking an arrow on the phase line. To check the direction of the arrows, it is important to consider the stability of each fixed point. If a fixed point is unstable, with $f'(x_0) > 0$, the arrows should point away from the fixed point, but near a stable fixed point ($f'(x_0) < 0$) the arrows should point towards the fixed point.

The phase line is very useful, since it shows the qualitative behaviour of the differential equation for any choice of initial condition.

Example 3.5 Find the fixed points, determine their stability, and sketch the phase line, for

$$\frac{dx}{dt} = x^2 - x^4. \tag{3.15}$$

The right-hand side can be factorised to give $x^2(1 - x^2) = x^2(1 - x)(1 + x)$. The fixed points are therefore $x = -1$, $x = 0$ and $x = 1$. From (3.15) it is clear that $\dot{x} > 0$ if $-1 < x < 0$ or $0 < x < 1$, so in these regions x increases and the arrow

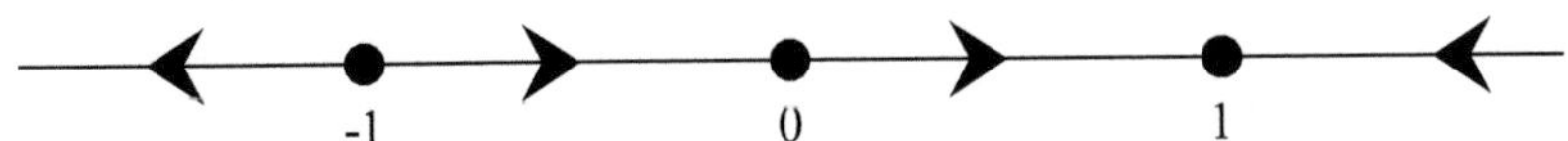

Fig. 3.2 Phase line for the differential equation (3.15), Example 3.5

on the phase line is drawn pointing to the right. But for $x < -1$ or $x > 1$, $\dot{x} < 0$ so
the arrows must point to the left. Hence the phase line is as shown in Fig. 3.2.

To calculate the stability of the fixed points, set $f(x) = x^2 - x^4$, find $f'(x) = 2x - 4x^3$ and evaluate $f'(x)$ at each fixed point. At $x = -1$, $f'(x) = 2$, so this point is unstable. $f'(0) = 0$, so $x = 0$ is a borderline case, as described in Sect. 3.2.2. Near $x = 0$, $f(x) > 0$, so x is increasing. Since solutions with $x > 0$ move away from $x = 0$, the fixed point $x = 0$ is classified as unstable. Finally, $f'(1) = -2$ so the fixed point at $x = 1$ is stable.

Hence the arrows on the phase line are consistent with the stability of each fixed point. The arrows point towards the stable fixed point at $x = 1$ and away from the unstable one at $x = -1$, as expected.

Having sketched the phase line, we can now describe the solution of (3.15) for any choice of initial condition. For example, if $x(0) = 2$, $x(t)$ decreases and approaches the point $x = 1$ as $t \to \infty$. If $x(0) = -0.9$, $x(t)$ increases and approaches the point $x = 0$ as $t \to \infty$.

3.3.2 Solution Curves

Having obtained the phase line, we can draw qualitative solution curves to the differential equation, showing $x(t)$. There are infinitely many possible solution curves, corresponding to different choices of initial condition. The simplest solution curves correspond to the fixed points of the equation; these are horizontal lines on the graph since x remains constant at a fixed point. Solution curves can then be drawn for initial conditions near a fixed point. If the fixed point is stable, then nearby solutions approach it, in the form of an exponentially decaying curve. For an unstable fixed point, nearby solutions move away exponentially initially, but the exponential behaviour is only seen when the perturbation is small, so when the solution has moved away significantly from the fixed point, the curve is no longer exponential.

When sketching solution curves, we can make use of the fact that they do not cross each other. If two solution curves did cross, they would have different slopes at that point, but (3.2) specifies the slope of the curve uniquely, so this cannot happen.

Example 3.6 Sketch the phase line and solution curves for the differential equation

$$\frac{dx}{dt} = 2\sin x - x. \tag{3.16}$$

Although this equation is separable, the integral can't be done in terms of standard functions, so unlike Example 3.1, we can't write down an analytical solution.

First we find the fixed points of the equation, values of x for which $f(x) = 2\sin x - x = 0$. One fixed point is clearly $x = 0$. But are there others? At the point $x = \pi/2$, $f(x) = 2 - \pi/2 > 0$, and at $x = \pi$, $f(x) = -\pi < 0$, so by the intermediate value theorem, there must be a fixed point somewhere between $x = \pi/2$ and $x = \pi$, and since $f'(x) < 0$ in this interval, there is only one fixed point in this interval. We will call this point $x = a$. Also, since the function $f(x)$ is an odd function (recall that this means that $f(-x) = -f(x)$), it must also be the case that $x = -a$ is a fixed point. Therefore, there are three fixed points, $x = -a$, $x = 0$ and $x = a$. There are no more fixed points, because for $x > \pi$, $f(x) < 2 - \pi < 0$.

Now consider the stability of the three fixed points. This is determined by

$$f'(x) = 2\cos x - 1.$$

For the fixed point at $x = 0$, $f'(0) = 1$, so this fixed point is unstable. For the fixed point at $x = a$, $f'(a) = 2\cos a - 1$. Since it is known that $\pi/2 < a < \pi$, it follows that $\cos a < 0$ so $f'(a) < 0$ and the point at $x = a$ is stable. Similarly, the point at $x = -a$ is stable. Using the stability of the fixed points, we can now plot the phase line, see Fig. 3.3.

Using the phase line, we can sketch the solution curves $x(t)$ to the differential equation (3.16). These are shown in Fig. 3.4 and the different types of solution are described below:

- There are three constant solutions, corresponding to the three fixed points, $x = -a$, $x = 0$ and $x = a$.
- A solution curve that starts from an initial condition $x(0) = 3$, which is greater than a, decreases and approaches the stable fixed point $x = a$ as $t \to \infty$. The final approach to the fixed point is in the form of an exponentially decaying curve.
- With an initial condition $x(0) = 1$, x increases and again approaches the point $x = a$.
- If the initial condition is $x(0) = 0.1$, close to the unstable fixed point at $x = 0$, the solution starts to grow exponentially away from $x = 0$, and then again approaches $x = a$ as $t \to \infty$.
- Because of the symmetry of the equation, the solutions with initial conditions $x(0) < 0$ are a mirror image of those with $x(0) > 0$.

In this example, any solution starting from a positive initial condition approaches the stable fixed point at $x = a$. This range of initial conditions is sometimes known as the *domain of stability* or the *basin of attraction* of the fixed point. Similarly, all negative initial conditions lie in the basin of attraction of the fixed point at $x = -a$.

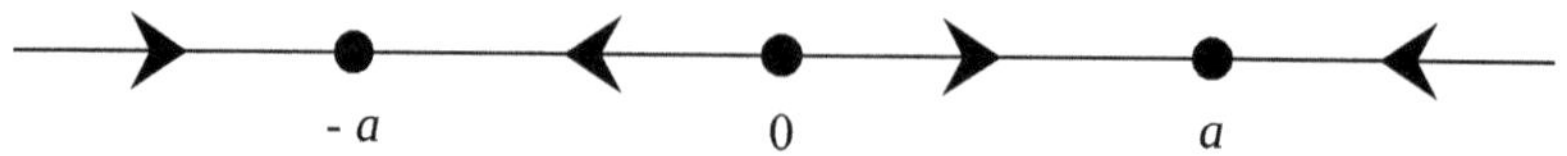

Fig. 3.3 Phase line for the differential equation (3.16)

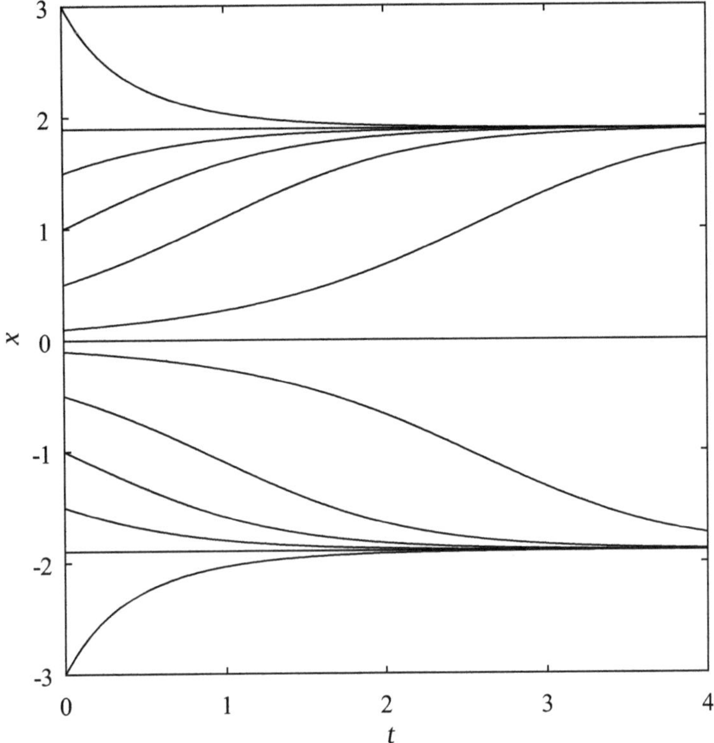

Fig. 3.4 Solution curves for the differential equation (3.16)

3.4   Parameters and Bifurcations

In many applications, a differential equation includes one or more parameters, and we would like to be able to describe the qualitative behaviour of the equation for different values of the parameters. As the value of a parameter changes, the qualitative picture as shown by the phase line may remain the same. However, there may be certain parameter values at which a significant change occurs. For example, the number of fixed points might change, or a stable fixed point may change into an unstable one. When this occurs, the equation is said to have *bifurcation*. Bifurcations will be considered in detail in Chap. 6, but here is an introductory example.

Example 3.7   Suppose that a parameter μ is introduced to (3.15) so that

$$\frac{dx}{dt} = \mu + x^2 - x^4. \tag{3.17}$$

Investigate how the number of fixed points depends on μ and hence deduce the values of μ at which bifurcations occur.

The equation for a fixed point is $x^4 - x^2 - \mu = 0$, which is a quadratic for x^2. There are two real roots for x^2 if $1 + 4\mu > 0$, so $\mu > -1/4$. These two roots are both positive, giving four real solutions for x, if $\mu < 0$. If $\mu > 0$, one value for x^2 is positive and the other is negative, so there are two real fixed points. For the case $\mu = 0$ there are three solutions, as shown in Fig. 3.2, and in the other borderline case, $\mu = -1/4$, the fixed-point equation is $(x^2 - 1/2)^2 = 0$ so there are two fixed points. For $\mu < -1/4$ there are no fixed points. There are bifurcations at these borderline values of μ where the number of solutions changes, so bifurcations occur at $\mu = 0$ and $\mu = -1/4$.

3.5 Qualitative Methods for Non-Autonomous Equations

Up to this point, it has been assumed that the differential equation that we are trying to describe qualitatively is autonomous. That is, the independent variable t does not appear explicitly in the equation. In that case, the essential behaviour of the differential equation can be described by a simple one-dimensional diagram, the phase line.

This section describes the qualitative approach to a first-order differential equation that is not autonomous,

$$\frac{dx}{dt} = f(x, t). \tag{3.18}$$

Since t now appears in the equation, a one-dimensional phase line won't work, but we can still sketch solutions to the differential equation using a diagram called the *direction field* or *slope field*.

The direction field is a graph of x against t, that shows the direction, or slope, of solution curves at a grid of points. This makes use of the fact that (3.18) gives the direction or the slope of the solutions at any point in the (x, t) plane. To construct the direction field:

- Choose a point (x, t) in the plane and evaluate $f(x, t)$ to find the slope of solutions at that point.
- Draw a short line segment at this point, with the correct slope.
- Repeat this for a grid of values for x and t to build up a picture of line elements.
- Finally, draw solution curves by sketching smooth curves that follow the direction of these line elements.

This method is best illustrated by an example.

Example 3.8 Sketch the direction field for the differential equation

$$\frac{dx}{dt} = t - x, \tag{3.19}$$

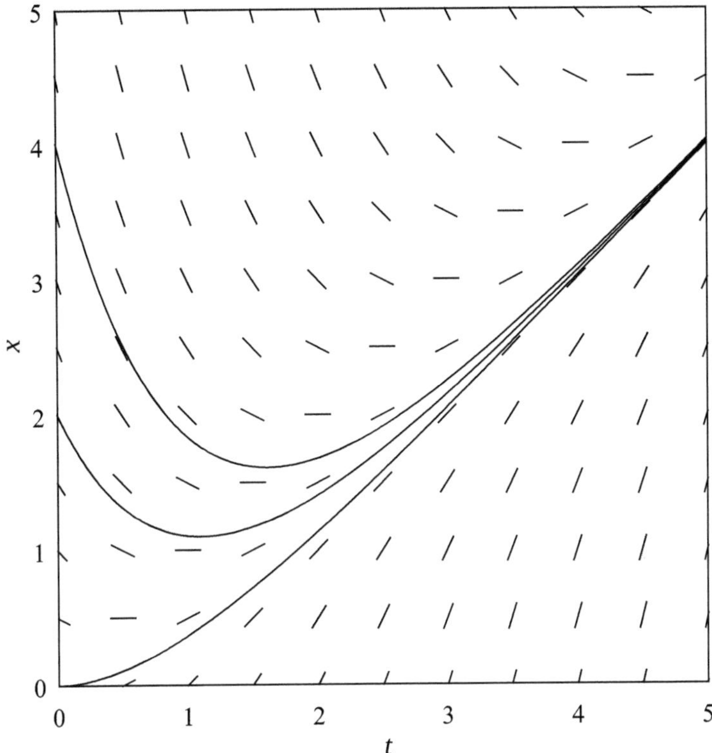

Fig. 3.5 Direction field and solution curves for the differential equation (3.19)

for $x > 0, t > 0$, and hence sketch solution curves for the initial conditions $x(0) = 0$, $x(0) = 2$ and $x(0) = 4$. What does the sketch suggest about the behaviour of solutions for large t?

The slope of the curve $x(t)$ is equal to $t - x$. For example, at the point $t = 1$, $x = 1$, the slope is zero, so we draw a short line with zero slope at this point, in other words, a horizontal line. Similarly, at all points where $t = x$, the slope is zero. At the point $t = 0$, $x = 1$, the slope is -1, so at this point we draw a line segment with slope -1, and the slope is the same at all points along the line $x = t + 1$ in the plane. By continuing this process, we can build up a grid of line segments indicating the slope or direction of solutions (Fig. 3.5).

Having drawn the direction field, we can now sketch solution curves, starting from each of the three initial conditions given. The curves are drawn so that they always follow the directions given by the line segments in the direction field, as shown in Fig. 3.5.

The sketched solutions show that $x \to \infty$ as $t \to \infty$. Furthermore, the solution curves all seem to approach the line $x = t - 1$. It is easy to check that this line $x = t - 1$ is a solution of (3.19). In fact, (3.19) can be solved analytically using the methods of the previous chapter, see Exercise 3.9.

Key Points from Chap. 3

- Many differential equations cannot be solved analytically. In these cases, a *qualitative* approach is often helpful for understanding the behaviour of solutions of the equation.
- A *qualitative* method means describing and sketching the solution to a differential equation, without obtaining an analytical solution.
- A first-order differential equation is *autonomous* if it does not depend explicitly on the independent variable, that is, it can be written in the form $\dot{x} = f(x)$.
- For the first-order autonomous differential equation $\dot{x} = f(x)$, a *fixed point* is a value x_0 such that $f(x_0) = 0$.
- Near a fixed point, the differential equation may be *linearised*, using a truncated Taylor expansion about the point $x = x_0$, usually leading to exponentially decaying or growing solutions.
- A fixed point x_0 is *stable* if $f'(x_0) < 0$, meaning that nearby solutions are attracted towards it.
- A fixed point with $f'(x_0) > 0$ is *unstable*, which means that nearby solutions move away from it.
- In the borderline case $f'(x_0) = 0$, the linearisation is not valid and higher order terms need to be included in the Taylor expansion about x_0.
- The behaviour of a first-order autonomous differential equation can be represented qualitatively on a *phase line*, showing the fixed points and arrows between the fixed points indicating whether $x(t)$ is increasing or decreasing.
- Using the phase line, we can sketch solution curves $x(t)$, for any choice of initial condition.
- A differential equation including a parameter has a *bifurcation* at a parameter value where the number of fixed points changes or the stability of a fixed point changes.
- For the non-autonomous equation $\dot{x} = f(x, t)$, the phase line method is not applicable, but we can sketch the direction field and hence sketch solution curves $x(t)$.

Exercises

3.1 Show that the differential equation

$$\frac{dx}{dt} = x^2 - x - 6$$

has two fixed points, and find which one is stable.

3.2 Construct a differential equation that has stable fixed points at $x = \pm 4$ and an unstable fixed point at $x = 1$.

3.3 A modification of the logistic model of population growth (3.9) is

$$\frac{dx}{dt} = ax - bx^3,$$

where a and b are positive constants.

(a) Following the method of Example 3.4, nondimensionalise the equation to obtain a form with no constants.
(b) Find the fixed points of the dimensionless equation and determine whether they are stable or unstable. Hence describe the long-term behaviour of solutions for an initial condition $x(0) > 0$.
(c) Find the analytical solution of the dimensionless equation using an appropriate method from Chap. 2 and check that this solution is consistent with your description in part (b).

3.4 The downward vertical velocity v of a particle falling through the Earth's atmosphere with air resistance proportional to v^2 obeys the differential equation

$$\frac{dv}{dt} = g - kv^2,$$

where g is the acceleration due to gravity and $k > 0$ is a constant.

(a) Without solving the equation, describe how v changes with time if the particle starts from rest.
(b) By considering the phase line, or otherwise, find a flaw in this model.
(c) How could this flaw be corrected?

3.5 (a) Find the solution of (3.13), with the initial condition $y(0) = y_0$.
(b) In the stable case, describe how solutions approach $y = 0$ as $t \to \infty$.
(c) Describe the behaviour in the unstable case.

3.6 For each of the following autonomous differential equations, find all the fixed points, if possible; otherwise, find how many fixed points there are. Investigate the stability of each fixed point, sketch the phase line and describe the solutions.

(a)

$$\frac{dx}{dt} = x^4 - 5x^2 + 4,$$

(b)

$$\frac{dx}{dt} = x^3 - 3x - 1,$$

(c)

$$\frac{dx}{dt} = xe^x - x,$$

(d)

$$\frac{dx}{dt} = \sin(x^2).$$

3.7 The differential equation $\dot{x} = f(x)$ is symmetrical under the reflection $x \leftrightarrow -x$. That is, replacing x by $-x$ gives the same equation. Show that $x = 0$ is a fixed point. What can be deduced if the symmetry is $x \leftrightarrow c - x$?

3.8 Describe how the existence of fixed points of the differential equation

$$\frac{dx}{dt} = \mu x + x^3$$

depends on the parameter μ. Hence find the value of μ at which a bifurcation occurs. Describe the behaviour of solutions of the differential equation for all values of μ.

3.9 Using the methods of the previous chapter, find the analytical solution of (3.19), and hence confirm that solutions approach the line $x = t - 1$ as $t \to \infty$.

3.10 (a) Sketch the direction field for the differential equation

$$\frac{dx}{dt} = tx^2,$$

for $t > 0$, including the regions $x > 0$ and $x < 0$.
(b) Hence sketch and describe the behaviour of solutions with initial conditions $x(0) > 0$ and $x(0) < 0$.
(c) Find the solution of the differential equation and check that it is consistent with your direction field sketch.

Chapter 4
Second-Order Linear Systems

The aim of this chapter and the next is to extend the idea of a qualitative solution to second-order differential equations and systems of two coupled first-order equations. This chapter discusses linear systems, and Chap. 5 considers the nonlinear case. The behaviour of the system can be described using a *phase plane*, analogous to the phase line used in the previous chapter.

A general linear, second-order, autonomous differential equation for $u(t)$ has the form

$$a_2\ddot{u} + a_1\dot{u} + a_0 u = k,$$

where a_2, a_1, a_0 and k are constants. This can be written as a homogeneous equation

$$a_2\ddot{x} + a_1\dot{x} + a_0 x = 0,$$

by changing variable from u to $x = u - k/a_0$ (if $a_0 = 0$, the equation can be regarded as a first-order one for $\dot{u}$).

This second-order equation can be written as two first-order equations for $x(t)$ and a new variable $y(t)$ simply by defining $y = \dot{x}$, so that

$$\dot{x} = y, \qquad a_2\dot{y} + a_1 y + a_0 x = 0.$$

Recall that the reverse process, writing two coupled first-order differential equations as one second-order one, was described in Sect. 2.4. As discussed there, this linear system can be solved analytically by eliminating one of the variables. But the aim of this chapter is to give a neater method of solution and to introduce a geometrical, qualitative view of the behaviour. This will be very useful when considering the nonlinear case in the next chapter.

© The Author(s), under exclusive license to Springer Nature Switzerland AG 2025   39
P. C. Matthews, *Differential Equations, Bifurcations and Chaos*,
Springer Undergraduate Mathematics Series,
https://doi.org/10.1007/978-3-031-99543-9_4

One of the remarkable things about mathematics is how topics that appear to be unrelated to each other are in fact connected. A good example of this is the link between differential equations and linear algebra. The Appendix includes a summary of the topic of eigenvalues and eigenvectors of matrices that will be used throughout this chapter and the remainder of this book.

4.1 Eigenvalues, Eigenvectors and Differential Equations

What do matrices, eigenvalues and eigenvectors have to do with differential equations? It turns out that for systems of coupled, linear differential equations, eigenvalues and eigenvectors arise naturally when we seek a solution.

Consider the autonomous linear system

$$\dot{x} = ax + by,$$
$$\dot{y} = cx + dy. \tag{4.1}$$

Section 2.5 showed how to solve this type of linear system by differentiating one of the equations with respect to t, and substituting twice to eliminate one of the variables. This leads to a single second-order differential equation that can be solved by seeking exponential solutions.

But since we know that the solutions are exponential, a better approach is to seek an exponential solution immediately. There is no need to eliminate one of the variables. First, we can write the system (4.1) in matrix form as

$$\begin{pmatrix} \dot{x} \\ \dot{y} \end{pmatrix} = \begin{pmatrix} a & b \\ c & d \end{pmatrix} \begin{pmatrix} x \\ y \end{pmatrix}. \tag{4.2}$$

Now we try an exponential solution, setting $x = x_0 e^{\lambda t}$, $y = y_0 e^{\lambda t}$ in (4.2). Then, since $\dot{x} = \lambda x$ and $\dot{y} = \lambda y$, this leads to

$$\lambda \begin{pmatrix} x_0 \\ y_0 \end{pmatrix} = \begin{pmatrix} a & b \\ c & d \end{pmatrix} \begin{pmatrix} x_0 \\ y_0 \end{pmatrix}. \tag{4.3}$$

which can be written as

$$\lambda v = M v \quad \text{where} \quad v = \begin{pmatrix} x_0 \\ y_0 \end{pmatrix} \quad \text{and} \quad M = \begin{pmatrix} a & b \\ c & d \end{pmatrix}. \tag{4.4}$$

This is exactly Eq. (A.4) defining λ as an eigenvalue of M and v as an eigenvector. So the growth rate λ of an exponential solution of a system of linear differential equations is an eigenvalue of the matrix of coefficients.

Assuming that there are two distinct eigenvalues, λ_1 and λ_2, with corresponding eigenvectors $(x_1, y_1)^T$ and $(x_2, y_2)^T$, there are two exponential solutions,

$$\begin{pmatrix} x \\ y \end{pmatrix} = e^{\lambda_1 t} \begin{pmatrix} x_1 \\ y_1 \end{pmatrix} \quad \text{and} \quad \begin{pmatrix} x \\ y \end{pmatrix} = e^{\lambda_2 t} \begin{pmatrix} x_2 \\ y_2 \end{pmatrix}.$$

Using the superposition principle for linear, homogeneous equations, the general solution is a linear combination of these two solutions,

$$\begin{pmatrix} x \\ y \end{pmatrix} = A e^{\lambda_1 t} \begin{pmatrix} x_1 \\ y_1 \end{pmatrix} + B e^{\lambda_2 t} \begin{pmatrix} x_2 \\ y_2 \end{pmatrix}, \tag{4.5}$$

where A and B are arbitrary constants.

Example 4.1 Find the general solution of the system

$$\begin{aligned} \dot{x} &= 2x + y, \\ \dot{y} &= x + 2y, \end{aligned} \tag{4.6}$$

and find the solution that obeys the initial conditions $x(0) = 4$, $y(0) = 0$.
   In matrix form,

$$\begin{pmatrix} \dot{x} \\ \dot{y} \end{pmatrix} = \begin{pmatrix} 2 & 1 \\ 1 & 2 \end{pmatrix} \begin{pmatrix} x \\ y \end{pmatrix},$$

and seeking exponential solutions $x = x_0 e^{\lambda t}$, $y = y_0 e^{\lambda t}$ leads to

$$\lambda v = M v \quad \text{where} \quad v = \begin{pmatrix} x_0 \\ y_0 \end{pmatrix} \quad \text{and} \quad M = \begin{pmatrix} 2 & 1 \\ 1 & 2 \end{pmatrix}.$$

This shows that λ must be an eigenvalue of M. The eigenvalues of M are found using (A.6) as follows:

$$\begin{vmatrix} 2 - \lambda & 1 \\ 1 & 2 - \lambda \end{vmatrix} = 0 \Rightarrow (2 - \lambda)^2 - 1 = 0 \Rightarrow 2 - \lambda = \pm 1 \Rightarrow \lambda = 1 \text{ or } \lambda = 3.$$

In this example, there is no need to expand the bracket $(2 - \lambda)^2$ and then solve a quadratic equation; just move the 1 to the other side of the equation, take the square root and solve for the two values of λ. This short-cut works when the two diagonal entries in the matrix are equal. Finding the eigenvectors as in Example A.3,

$$\lambda = 1 \text{ has eigenvector } \begin{pmatrix} 1 \\ -1 \end{pmatrix} \quad \text{and} \quad \lambda = 3 \text{ has eigenvector } \begin{pmatrix} 1 \\ 1 \end{pmatrix}.$$

Hence there are two possible exponential solutions of the system,

$$\begin{pmatrix} x \\ y \end{pmatrix} = e^t \begin{pmatrix} 1 \\ -1 \end{pmatrix} \quad \text{and} \quad \begin{pmatrix} x \\ y \end{pmatrix} = e^{3t} \begin{pmatrix} 1 \\ 1 \end{pmatrix}.$$

The general solution is a linear combination of these two solutions,

$$\begin{pmatrix} x \\ y \end{pmatrix} = A e^t \begin{pmatrix} 1 \\ -1 \end{pmatrix} + B e^{3t} \begin{pmatrix} 1 \\ 1 \end{pmatrix}, \tag{4.7}$$

where A and B are arbitrary constants, or equivalently

$$x = A e^t + B e^{3t}, \qquad y = -A e^t + B e^{3t}.$$

For the initial conditions $x(0) = 4$, $y(0) = 0$, A and B must obey $A + B = 4$, $-A + B = 0$, so $A = B = 2$ and the solution is

$$x = 2(e^t + e^{3t}), \qquad y = 2(-e^t + e^{3t}).$$

4.2 Phase Planes and Nullclines

The behaviour of the linear system (4.2) can be represented by a plot of solutions in the x, y plane called a *phase plane* or *phase portrait*. Solutions move through the phase plane as the independent variable t increases, in a direction indicated by arrows, forming curves in the phase plane known as *trajectories*. The phase plane is the natural extension to two dimensions of the phase line for a first-order equation (see Sect. 3.3.1). The direction of a trajectory is clearly defined at all points of the plane for (4.2), except at the origin, which is a *fixed point* of (4.2). Fixed points are defined by $\dot{x} = \dot{y} = 0$, which again is a natural extension of the first-order case defined in Definition 3.1. The origin is the only fixed point of (4.2).

A slow way to sketch the phase plane is to choose a point (x, y) in the plane, evaluate the vector $(\dot{x}, \dot{y})$ at that point, and then draw an arrow, at the point (x, y) in the direction of this vector. After this has been done for several points in the plane, solution trajectories can be drawn as smooth curves following the direction of these arrows. This method is similar to drawing the direction field for a non-autonomous first-order equation (see Sect. 3.5).

A useful tool for drawing phase planes is the idea of a *nullcline*. The line in the plane on which $\dot{x} = 0$ is known as the *x-nullcline*. For example, for the linear system (4.2), the x-nullcline is the line $ax + by = 0$. On this line, since $\dot{x} = 0$, the direction of the trajectories is only in the y direction, so the arrows on the phase plane diagram point either upwards or downwards, depending on the sign of $\dot{y}$. Similarly, the line on which $\dot{y} = 0$ is called the *y-nullcline*, and on this line the trajectories all point in a horizontal direction, to the left or to the right, depending on the sign of $\dot{x}$.

Another important rule that is helpful for drawing phase planes is that two trajectories cannot cross, because (4.2) defines the vector giving the direction of a trajectory at all points, except at the fixed point at the origin where the vector has zero magnitude. Therefore trajectories cannot cross each other, and can only meet at the origin.

Example 4.2 Sketch the nullclines and the phase plane for the system (4.6).

In (4.6), $\dot{x} = 2x + y$, so the *x*-nullcline is the line $y = -2x$. On this line, $\dot{x} = 0$, so the trajectories point in the *y* direction, either up or down. To determine which, consider $\dot{y} = x + 2y$. On the *x*-nullcline, $\dot{y} = -3x$, so if $x < 0$, $\dot{y} > 0$ and trajectories point upwards, but when $x > 0$, $\dot{y} < 0$ so arrows on the phase plane point downwards.

Similarly, the *y*-nullcline, where $\dot{y} = 0$, is the line $y = -x/2$. On this line all trajectories point horizontally, left or right, and $\dot{x} = 3x/2$ so $\dot{x} > 0$ and arrows point to the right if $x > 0$, but arrows point to the left if $x < 0$. The two nullclines are shown as dashed lines in Fig. 4.1, together with vectors indicating the direction of trajectories as they cross these lines.

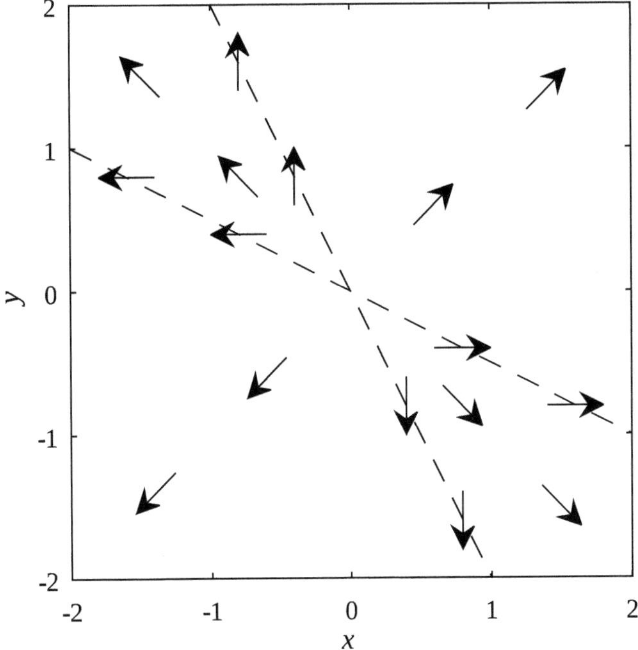

Fig. 4.1 Partial phase plane for Example 4.2. The *x*-nullcline, $y = -2x$, and *y*-nullcline, $y = -x/2$, are marked as dashed lines. Arrows show the direction of solution trajectories on both the nullclines and on the eigenvector lines $y = x$ and $y = -x$

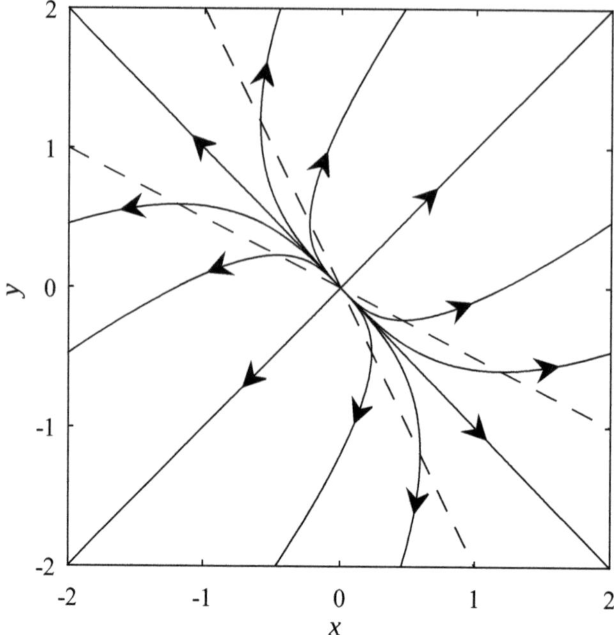

Fig. 4.2 Full phase plane for Example 4.2. As Fig. 4.1, but including several trajectories of the system

At any point in the plane lying on an eigenvector, the vector $(\dot{x}, \dot{y})$ is parallel to the vector (x, y). In this example, one eigenvector is $(1, 1)^T$, so at points where $y = x$, $\dot{y} = \dot{x}$. Checking the signs of $\dot{x}$ and $\dot{y}$ shows that trajectories point up and to the right ($\dot{x} = \dot{y} > 0$) when $y = x > 0$ and down and to the left when $y = x < 0$. Similarly, at points where $y = -x$, on the eigenvector $(1, -1)^T$, trajectories point up and to the left if $y = -x > 0$, down and to the right if $y = -x < 0$, as shown in Fig. 4.1.

Using these arrows on the nullclines and eigenvectors, smooth curves of solution trajectories can be drawn, see Fig. 4.2. These are straight lines for solutions lying exactly on the eigenvectors, corresponding to $A = 0$ or $B = 0$ in (4.7) but otherwise they are curved. As t increases, the e^{3t} becomes the dominant term in (4.7), and this means that solutions become more aligned with the eigenvector $y = x$.

4.3   Classification of Linear Systems

The aim of this section is to classify different types of behaviour of the linear system of two coupled first-order, autonomous differential equations (4.1), which can be written in matrix form as

$$\begin{pmatrix} \dot{x} \\ \dot{y} \end{pmatrix} = \begin{pmatrix} a & b \\ c & d \end{pmatrix} \begin{pmatrix} x \\ y \end{pmatrix}. \tag{4.8}$$

As shown in Sect. 4.1, the solutions of (4.8) are exponential, with possible exponential growth rates λ given by the eigenvalues of the matrix of the linear system.

For a general matrix

$$M = \begin{pmatrix} a & b \\ c & d \end{pmatrix}$$

the eigenvalues are found from the characteristic equation

$$|M - \lambda I| = 0 \Rightarrow \begin{vmatrix} a - \lambda & b \\ c & d - \lambda \end{vmatrix} = 0 \Rightarrow \lambda^2 - (a+d)\lambda + ad - bc = 0.$$

This can be written as

$$\lambda^2 - T\lambda + D = 0 \tag{4.9}$$

where T is the trace of the matrix and D is its determinant, defined by

$$T = \text{Trace}(M) = a + d, \qquad D = |M| = ad - bc.$$

Using the rules for sums and products of roots of quadratic equations, this shows that the two eigenvalues λ_1 and λ_2 obey

$$\lambda_1 + \lambda_2 = T, \qquad \lambda_1 \lambda_2 = D.$$

The qualitative behaviour of the linear system (4.8) is determined by the eigenvalues of the matrix. There are three main types of behaviour, corresponding to real eigenvalues of the same sign, real eigenvalues of opposite sign, and complex eigenvalues, and a number of sub-cases and special cases, described in detail in the following sections. Each type of behaviour can be illustrated with a distinctive phase portrait in the x, y plane.

4.3.1   Nodes: Two Real Eigenvalues with the Same Sign

Consider first the case where the two solutions λ_1, λ_2 of (4.9) are real, distinct, and have the same sign. This occurs if

$$T^2 > 4D \quad \text{and} \quad D > 0.$$

The first inequality is the condition for the quadratic (4.9) to have two real roots. The second one comes from the fact that the determinant is the product of the two roots, so it must be positive if the roots have the same sign.

The general solution (4.5) consists of a linear combination of exponentials. If both eigenvalues are positive, as in Example 4.1, then the system shows exponential growth and it is referred to as an *unstable node*. But if the two eigenvalues are negative, solutions decay exponentially to zero, and this case is known as a *stable node*.

To sketch the solutions in the phase plane, consider first the case $B = 0$ in (4.5), so that

$$\begin{pmatrix} x \\ y \end{pmatrix} = A e^{\lambda_1 t} \begin{pmatrix} x_1 \\ y_1 \end{pmatrix}.$$

In this case, $y/x = y_1/x_1$ for all t, so the solution is always aligned with the eigenvector, corresponding to a straight line in the x, y phase plane. Similarly, when $A = 0$ there is a solution that forms a straight line along the other eigenvector $(x_2, y_2)^T$.

However, for general values of A and B, the solution forms a curve, not a straight line, in the plane. Suppose that the eigenvalues of an unstable node are ordered so that $\lambda_1 > \lambda_2 > 0$ (for the special case $\lambda_1 = \lambda_2$, see Example 4.7 and Fig. 4.7). In this case, the first term in (4.5) grows more rapidly than the second, so as t increases, the solution curve becomes increasingly aligned with the first eigenvector, as shown in Fig. 4.2.

In the case of a stable node, suppose that $0 > \lambda_1 > \lambda_2$. Then the first term in (4.5) decays more slowly than the second, so again as t increases, the first term becomes increasingly large compared with the second, and again the trajectory becomes aligned with the first eigenvector, as shown in the following example.

Example 4.3 Classify the linear system

$$\dot{x} = -x + y,$$

$$\dot{y} = -2y,$$

and sketch its phase portrait.

The matrix of the system is

$$M = \begin{pmatrix} -1 & 1 \\ 0 & -2 \end{pmatrix},$$

with eigenvalues $\lambda_1 = -1$, $\lambda_2 = -2$. For a triangular matrix, the eigenvalues are just the diagonal entries of the matrix, so no calculation is needed. Since there are two real, negative eigenvalues, this system is a stable node.

To sketch the phase portrait, it is useful to find the eigenvectors. These are

$$v_1 = \begin{pmatrix} 1 \\ 0 \end{pmatrix} \text{ for } \lambda_1 = -1 \quad \text{and} \quad v_2 = \begin{pmatrix} 1 \\ -1 \end{pmatrix} \text{ for } \lambda_2 = -2.$$

The general solution is

$$\begin{pmatrix} x \\ y \end{pmatrix} = A e^{-t} \begin{pmatrix} 1 \\ 0 \end{pmatrix} + B e^{-2t} \begin{pmatrix} 1 \\ -1 \end{pmatrix}.$$

Hence there are solution trajectories lying along the x axis, corresponding to the case $B = 0$, and solutions on the line $y = -x$, corresponding to $A = 0$. Both these straight-line solutions approach the origin exponentially. Since the sign of the eigenvector is arbitrary, there are two parts to each of these solutions, one each side of the origin. We can draw straight lines approaching the origin along the positive and negative x axis, and along the line $y = -x$ for $x > 0$ and for $x < 0$, as shown in Fig. 4.3. On each of these straight lines, arrows are drawn pointing towards the origin to show the change as t increases.

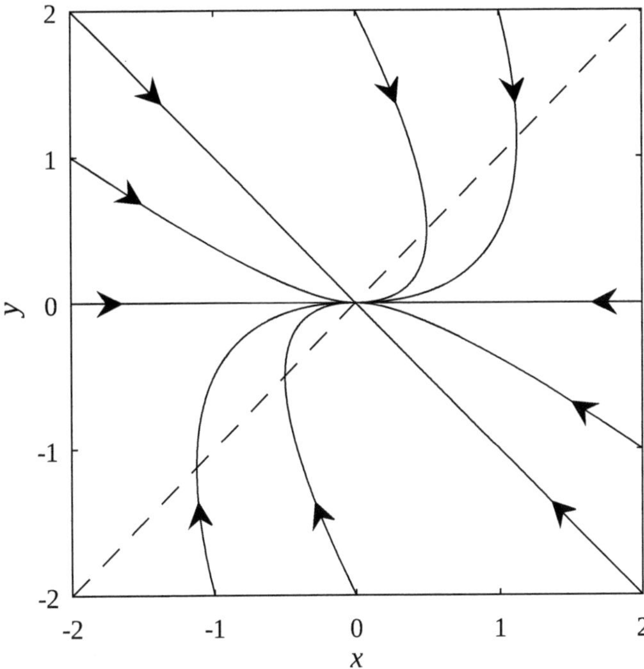

Fig. 4.3 Phase plane for Example 4.3, which is a stable node. Straight lines are solutions parallel to the eigenvectors v_1 and v_2. Other solutions are curved but become increasingly aligned with v_1 as they approach the origin. The dashed line is the x-nullcline, $y = x$. The y-nullcline is $y = 0$

Now consider the nullclines. The x-nullcline of the system is the line $y = x$, shown as a dashed line in Fig. 4.3. On this line $\dot{x} = 0$ and $\dot{y} = -2y$, so trajectories point in the negative y direction if $y > 0$ and in the positive y direction if $y < 0$. In the region of the plane below this nullcline, where $y < x$, $\dot{x} < 0$, and above it, $\dot{x} > 0$.

In this example the y-nullcline is the line $y = 0$ which corresponds to the eigenvector v_1. Above this line, $\dot{y} < 0$ and below it, $\dot{y} > 0$.

Finally, it is useful to consider the magnitude of the eigenvalues. Since $\lambda_2 < \lambda_1$, the contribution from the second term to the general solution (4.5) becomes very small compared with the first term as t increases, so trajectories curve in such a way as to align with the direction of the first term. Therefore, solutions curve so that they approach the origin closely aligned with the eigenvector v_1, which in this case is the x axis.

Putting all this information together gives the phase portrait shown in Fig. 4.3.

4.3.2   Saddles: Two Real Eigenvalues with Opposite Sign

If the two eigenvalues of a linear system are real but have opposite sign, then the system is described as a *saddle*. Since the determinant D is the product of the eigenvalues, this occurs if $D < 0$. If $D < 0$ then the quadratic (4.9) must have real roots, since $T^2 - 4D > 0$, so there is no need to check this condition.

In this case, there are solutions that approach the origin, decaying exponentially, following a path along the eigenvector v_- corresponding to the negative eigenvalue λ_-, and solutions that grow exponentially in the direction of the eigenvector v_+ of the positive eigenvalue λ_+. The general solution, with A and B both non-zero in (4.5), includes both growing and decaying exponentials. Solutions of this type form curved trajectories that become increasingly aligned with v_+ as $t \to \infty$, and align with v_- as $t \to -\infty$. To sketch the phase portrait, it is essential to calculate the eigenvectors, to find the unstable and stable directions.

Example 4.4   Classify the system

$$\dot{x} = x + y,$$

$$\dot{y} = 3x - y,$$

and sketch trajectories in the phase plane.

The matrix of the linear system is

$$M = \begin{pmatrix} 1 & 1 \\ 3 & -1 \end{pmatrix}, \tag{4.10}$$

with determinant $D = -4$, so we have eigenvalues of opposite sign, hence a saddle. The trace of (4.10) is $T = 0$, so using (4.9), the characteristic equation is

$$\lambda^2 - 4 = 0$$

so $\lambda = \pm 2$, and the eigenvectors are

$$v_+ = \begin{pmatrix} 1 \\ 1 \end{pmatrix} \text{ for } \lambda = \lambda_+ = 2 \quad \text{and} \quad v_- = \begin{pmatrix} 1 \\ -3 \end{pmatrix} \text{ for } \lambda = \lambda_- = -2.$$

To draw the phase plane, first consider the special case where only the exponentially growing solution is present. These solutions form a straight line along the line $y = x$, moving away from the origin in both directions. Similarly, there are straight line solutions along the line $y = -3x$ that approach the origin, marked by inward pointing arrows in the phase portrait, Fig. 4.4.

These straight lines divide the remainder of the phase portrait into four distinct regions, because trajectories cannot cross. In each of these regions, trajectories are

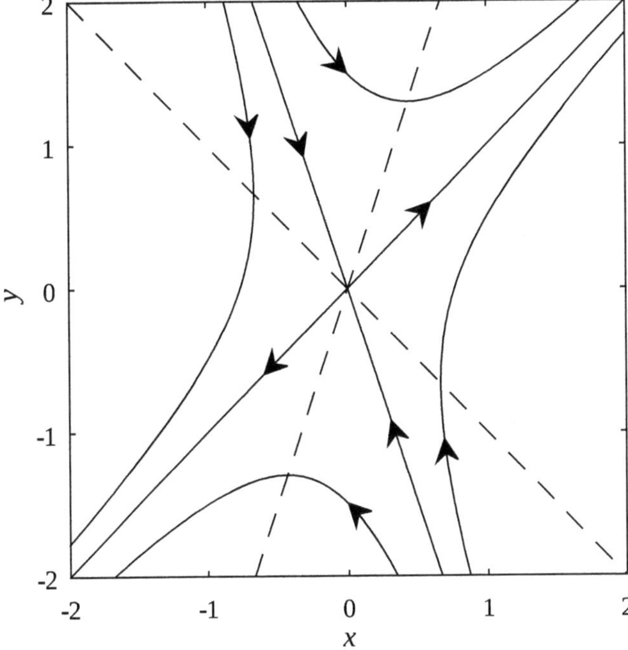

Fig. 4.4 Phase plane for Example 4.4 (saddle). The straight trajectories represent solutions parallel to the eigenvectors v_+ and v_-, corresponding to purely exponential growth or decay respectively. Other solutions are curved and become aligned with v_+ as t increases. Dashed lines are the x- and y-nullclines

curved lines that start near the v_- line and diverge from it, becoming more closely aligned with the v_+ direction as t increases, as shown in Fig. 4.4.

Again, the nullclines are helpful for sketching the phase portrait. The x-nullcline, which trajectories must cross vertically, is $y = -x$. The y-nullcline is $y = 3x$, and on this line, solution trajectories must be horizontal, as shown in Fig. 4.4.

4.3.3 Spirals: Complex Eigenvalues

The characteristic equation (4.9) that determines the eigenvalues of the linear system (4.8) has complex roots if $T^2 - 4D < 0$. In this case, the eigenvalues form a complex conjugate pair, $\lambda = \lambda_r \pm i\lambda_i$. The eigenvectors are also complex conjugates of each other. The general solution (4.5) then has the form

$$\begin{pmatrix} x \\ y \end{pmatrix} = Ae^{\lambda_r t + i\lambda_i t} \begin{pmatrix} x_1 \\ y_1 \end{pmatrix} + Be^{\lambda_r t - i\lambda_i t} \begin{pmatrix} \overline{x}_1 \\ \overline{y}_1 \end{pmatrix}, \tag{4.11}$$

where the bar indicates the complex conjugate. Since x and y must be real, $B = \overline{A}$ and (4.11) can be written as

$$\begin{pmatrix} x \\ y \end{pmatrix} = e^{\lambda_r t} \left(Ae^{i\lambda_i t} \begin{pmatrix} x_1 \\ y_1 \end{pmatrix} + \overline{A}e^{-i\lambda_i t} \begin{pmatrix} \overline{x}_1 \\ \overline{y}_1 \end{pmatrix} \right). \tag{4.12}$$

The interpretation of this solution is clearer if the complex numbers A, x_1, y_1 are written in modulus–argument form, $A = |A|e^{i\alpha}$, $x_1 = |x_1|e^{i\beta}$, $y_1 = |y_1|e^{i\gamma}$:

$$\begin{pmatrix} x \\ y \end{pmatrix} = 2e^{\lambda_r t}|A| \begin{pmatrix} |x_1|\cos(\alpha + \beta + \lambda_i t) \\ |y_1|\cos(\alpha + \gamma + \lambda_i t) \end{pmatrix}. \tag{4.13}$$

Solutions grow exponentially if $\lambda_r > 0$, or decay if $\lambda_r < 0$, and oscillate. The oscillations in x and y are out of phase, as long as $\beta \neq \gamma$, leading to solutions which rotate in the plane as they grow or decay. The case $\lambda_r > 0$ is known as an *unstable spiral*, while $\lambda_r < 0$ gives a *stable spiral*.

In examples, there is no need to calculate the complex eigenvectors. We find the sign of λ_r to see if the spiral is stable or unstable, and then find out whether the spiral rotates in a clockwise or anticlockwise sense, either by finding the direction of a trajectory at a specific point in the plane, or by considering the nullclines, as shown in the following example.

Example 4.5 Show that the system

$$\dot{x} = -x + 2y,$$

$$\dot{y} = -2x - y,$$

is a spiral. Determine whether the spiral is stable or unstable and whether the sense of rotation is clockwise or anti-clockwise, and sketch the phase portrait.

The characteristic equation for this system is

$$(-1 - \lambda)^2 + 4 = 0 \;\Rightarrow\; (1 + \lambda)^2 = -4 \;\Rightarrow\; 1 + \lambda = \pm 2i.$$

Therefore the eigenvalues are $\lambda = -1 \pm 2i$, which are complex, so the system describes a spiral. Alternatively, we can find the trace $T = -2$ and the determinant $D = 5$ and check that $T^2 - 4D < 0$. Since the real part of the eigenvalues is negative, this system is a stable spiral.

To investigate whether the spiral rotates in a clockwise or anti-clockwise sense, consider a point in the plane. For example, at $x = 1$, $y = 0$, $\dot{x} = -1$ and $\dot{y} = -2$, so the trajectory is directed downward and to the left. The sense of rotation is clockwise, since the direction is downward to the right of the origin. To sketch the phase plane, it is helpful to find the nullclines. The x-nullcline is $y = x/2$. On this line $\dot{x} = 0$ and $\dot{y} = -5x/2$, so trajectories point vertically downward when $x > 0$ and upward if $x < 0$. Similarly, the y-nullcline is $y = -2x$, and on this line trajectories point to the left if $x > 0$ and to the right if $x < 0$. Considering the directions of the trajectories on the nullclines confirms that trajectories rotate in a clockwise sense.

The phase plane is shown in Fig. 4.5. In this example, the two nullclines are perpendicular and the spiral has a symmetrical appearance. In general, this is not the case, and the spiral often appears to be 'squashed' in a particular direction, see Exercise 4.6.

4.3.4   Special Cases

The three main types of phase portrait for the linear system (4.8) are nodes, saddles and spirals, as discussed in the previous three sections. There are also several special cases, briefly described below.

If the trace and determinant obey the equation $T^2 - 4D = 0$, then there is only one solution to (4.9) and so only one eigenvalue. Usually, in this case there is also only one eigenvector. This case is called a *degenerate node*. It can be thought of as a borderline case between a node and a spiral. A special case of this special case is when the matrix of the linear system is a multiple of the identity matrix, which means that the two linear equations are uncoupled. In this case, any vector is an eigenvector, and trajectories are all straight lines moving out from, or towards, the origin. This is called a *star node*. Degenerate nodes and star nodes can be stable, if the single eigenvalue is negative, or unstable, if it is positive.

If the eigenvalues are both purely imaginary, solutions oscillate but do not grow or decay. This situation occurs if $T = 0$ and $D > 0$, and can be thought of as a borderline case between an unstable and stable spiral. The trajectories form closed ellipses in the phase plane, rotating in either a clockwise or anticlockwise sense.

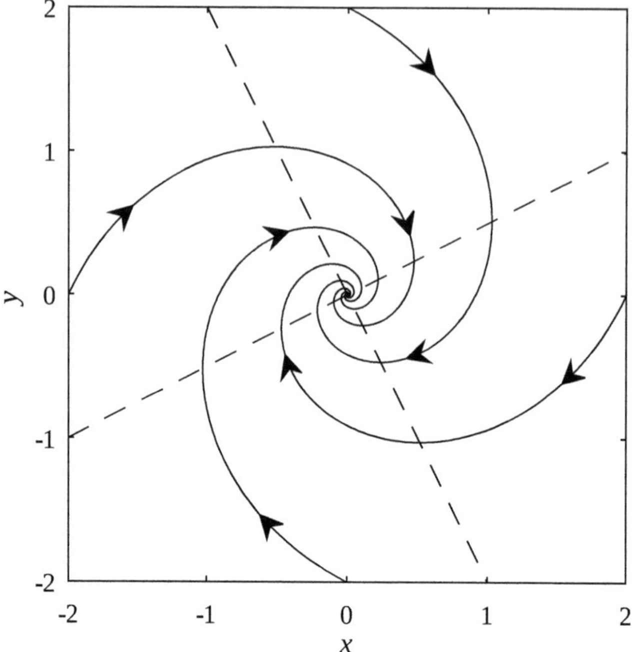

Fig. 4.5 Phase plane for Example 4.5 (stable spiral). Trajectories rotate clockwise and approach the origin. The dashed lines are the nullclines $y = x/2$ and $y = -2x$

This case is called a *centre*. The fact that trajectories are closed loops corresponds to a situation in which some quantity is conserved in the dynamics of the system. For example, centres are commonly found in physical systems where energy is conserved.

The following examples illustrate these special cases. There is another type of special case, where $D = 0$, which implies that one of the eigenvalues is zero. This case will be considered in Example 5.2 in Chap. 5.

Example 4.6 Find the type of the linear system

$$\dot{x} = 3x + y,$$
$$\dot{y} = -x + y,$$

and sketch the phase portrait.

The trace and determinant are $T = 4$, $D = 4$, so $T^2 = 4D$. Therefore the system is a degenerate node, with only one eigenvalue, $\lambda = 2$. Since this is positive, we have an unstable degenerate node. The eigenvector is $(1, -1)^T$. There are solution

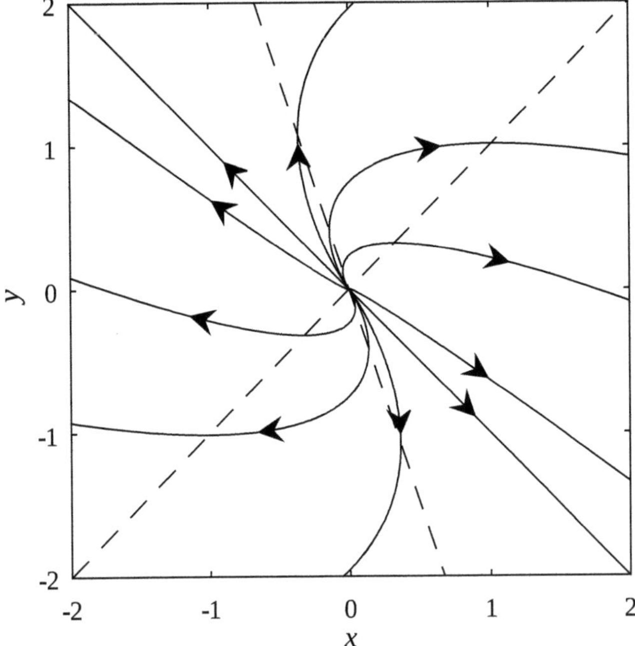

Fig. 4.6 Phase plane for the degenerate node, Example 4.6. The only straight trajectories are those along the eigenvector line $y = -x$. Dashed lines are the nullclines

trajectories along the line $y = -x$ of the form

$$\begin{pmatrix} x \\ y \end{pmatrix} = A e^{2t} \begin{pmatrix} 1 \\ -1 \end{pmatrix}.$$

To plot the phase portrait, it is useful to look at the nullclines. The x-nullcline is $y = -3x$, and trajectories cross this line vertically with $\dot{y} > 0$ if $x < 0$. The y-nullcline is $y = x$, and on this line $\dot{x} = 4x$ so trajectories point to the right if $x > 0$. It is also helpful to consider other points in the plane. For example, along the line $x = 0$, $\dot{x} = y = \dot{y}$, so trajectories cross this line diagonally, at an angle of $\pi/4$ relative to either axis. The phase plane is shown in Fig. 4.6.

Example 4.7 Classify the linear system

$$\dot{x} = -2x,$$
$$\dot{y} = -2y,$$

and sketch the phase portrait.

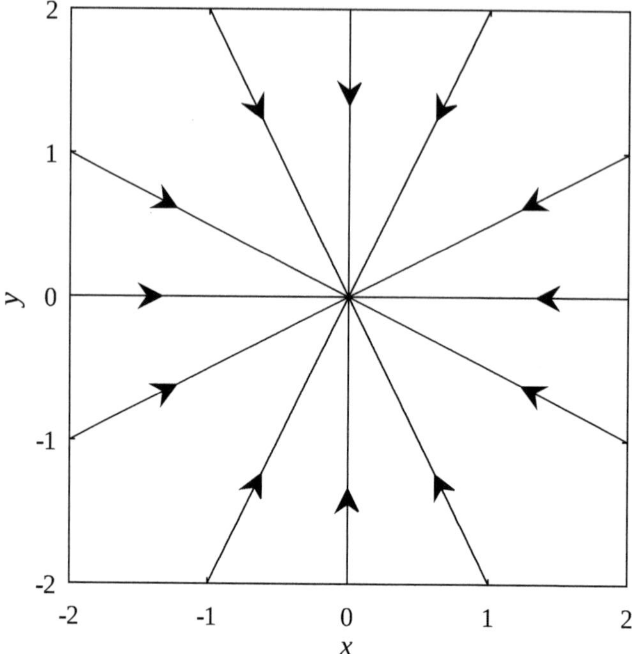

Fig. 4.7 Phase plane for the stable star node, Example 4.7

In this case, the two equations are uncoupled, and the matrix of the system is $-2I$. There is a single eigenvalue, $\lambda = -2$. Any vector v is an eigenvector, because $Iv = v$. The general solution can be written

$$\begin{pmatrix} x \\ y \end{pmatrix} = \mathrm{e}^{-2t} \begin{pmatrix} x_0 \\ y_0 \end{pmatrix}$$

for any constants x_0 and y_0, so on a trajectory, x/y takes the constant value x_0/y_0. Trajectories are therefore straight lines that approach the origin. The system is a stable star node, with a phase portrait as shown in Fig. 4.7.

Example 4.8 Show that the system

$$\dot{x} = 3x + 5y,$$

$$\dot{y} = -5x - 3y,$$

represents a centre. Find the direction of rotation, sketch the phase plane and find the equation of the ellipses that describe the trajectories.

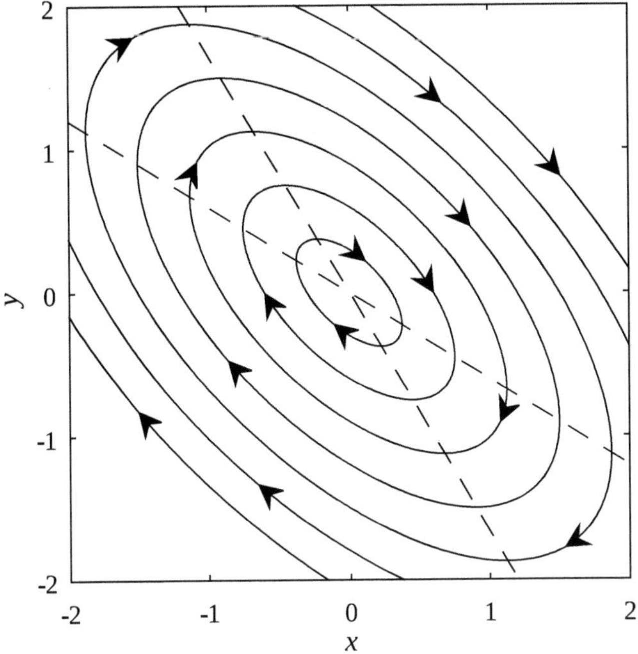

Fig. 4.8 Phase plane for the centre of Example 4.8. Dashed lines mark the x-nullcline $y = -3x/5$ and the y-nullcline $y = -5x/3$

The trace and determinant of the linear system are $T = 0$, $D = 16$, so the system is a centre, with purely imaginary eigenvalues, $\lambda = \pm 4i$. There is no need to find the complex eigenvectors. Finding the nullclines helps with determining the direction of rotation and sketching the phase plane. The x-nullcline is $y = -3x/5$, and on this line $\dot{x} = 0$ and $\dot{y} = -16x/5$. Since $\dot{y} < 0$ when $x > 0$, the direction of rotation is clockwise. Similarly, the y-nullcline is $y = -5x/3$, and on this line, $\dot{x} > 0$ when $y > 0$, so this confirms that the rotation is clockwise. Since the two nullclines are fairly close to each other, the trajectories have an elliptical shape, as shown in Fig. 4.8.

The general form of the equation of an ellipse with its centre at the origin is $Ax^2 + Bxy + Cy^2 = $ constant, where the constants A, B and C obey $B^2 - 4AC < 0$. If this quantity is constant on a trajectory, its derivative must be zero, so

$$2Ax\dot{x} + By\dot{x} + Bx\dot{y} + 2Cy\dot{y} = 0.$$

Substituting in $\dot{x}$ and $\dot{y}$ gives

$$(2Ax + By)(3x + 5y) + (Bx + 2Cy)(-5x - 3y) = 0,$$

and equating coefficients of x^2, xy and y^2 shows that

$$6A = 5B, \qquad A = C, \qquad 5B = 6C.$$

so the ellipses can be written in the form

$$5x^2 + 6xy + 5y^2 = c^2 = \text{constant}.$$

The ellipses have a reflection symmetry $x \leftrightarrow y$. The semi-major axis (the 'long' direction of the ellipse) lies on the line $y = -x$ with $4x^2 = c^2$, and the semi-minor axis is along $y = x$, where $16x^2 = c^2$. Hence the ratio of these two lengths is 2, which implies that the eccentricity of the ellipse is $\sqrt{3}/2$.

4.4   Stability and Instability

The concept of stability and instability for the linear system (4.8) has already been introduced in this chapter, through the terminology of, for example, the stable node, in which all solutions tend to the origin, and the unstable node, where solutions grow without limit. But what about the case of a centre: is it stable or unstable? The answer depends on the choice of definition of the word 'stable'. In this section, two slightly different definitions of stability are described.

Recall from Chap. 3 that stability is a property of a fixed point of a differential equation. For the linear system (4.8), the origin $x = 0$, $y = 0$ is a fixed point, since at that point $\dot{x}$ and $\dot{y}$ are both zero. This is the only fixed point, except in the special case when the determinant D of the matrix is zero.

The fixed point $(0, 0)$ is said to be *Lyapunov stable* if solutions that start near the origin remain near the origin for all time. The distance of a point $(x(t), y(t))$ from the origin is $d(t) = \sqrt{x(t)^2 + y(t)^2}$. The mathematical definition of Lyapunov stability is as follows.

Definition 4.1 The origin in (4.8) is *Lyapunov stable* if given any $\varepsilon > 0$, there exists a $\delta > 0$ such that if $d(0) < \delta$, then $d(t) < \varepsilon$ for all $t > 0$.

A stronger condition known as *asymptotic stability* holds if in addition, solutions that start sufficiently near the origin all approach the origin.

Definition 4.2 The origin in (4.8) is *asymptotically stable* if it is Lyapunov stable and there exists a $\delta > 0$ such that if $d(0) < \delta$, then $d(t) \to 0$ as $t \to \infty$.

If all the eigenvalues of a linear system have negative real part, then the origin is asymptotically stable. So stable nodes and stable spirals are asymptotically stable. If the origin is not Lyapunov stable, then it is said to be unstable. Unstable nodes and spirals, together with saddles, are unstable.

The borderline case which shows the difference between Lyapunov stable and asymptotically stable is the centre, where the eigenvalues have zero real part. A centre is Lyapunov stable, since solutions that start near the origin remain near the origin, but not asymptotically stable, since solutions do not approach the origin.

In Example 4.8, see Fig. 4.8, trajectories form ellipses with an aspect ratio of 2 between the semi-major and semi-minor axes. In this case, the distance from the origin may increase by a factor of at most 2, if the initial condition lies on the shorter axis. So to show Lyapunov stability, given any $\varepsilon > 0$, we can choose $\delta = \varepsilon/2$.

4.5 The Exponential of a Matrix

An alternative approach to solving the linear system (4.8) is to use the exponential of the matrix.

For a single first-order differential equation, $\dot{x} = ax$, with initial condition $x(0) = x_0$, the solution is an exponential function, $x = e^{at}x_0$. In a similar way, for a system of n linear equations, $\dot{u} = Mu$, where u is an $n \times 1$ vector and M is an $n \times n$ matrix, with initial condition $u(0) = u_0$, the solution can be written as $u = e^{Mt}u_0$. Here, the matrix exponential is defined similarly to the scalar exponential, by

$$e^{Mt} = I + Mt + \frac{M^2t^2}{2!} + \frac{M^3t^3}{3!}\cdots = \sum_{j=0}^{\infty} \frac{M^j t^j}{j!}. \tag{4.14}$$

From the definition, differentiating term by term, it can be seen that

$$\frac{d(e^{Mt})}{dt} = \sum_{j=0}^{\infty} \frac{M^j j t^{j-1}}{j!} = \sum_{j=1}^{\infty} \frac{M^j j t^{j-1}}{j!} = M \sum_{j=1}^{\infty} \frac{M^{j-1}t^{j-1}}{(j-1)!} = M e^{Mt},$$

where the last step follows after changing the index in the sum from j to $k = j - 1$. Hence $u = e^{Mt}u_0$ satisfies the system $\dot{u} = Mu$, and $e^{Mt} = I$ when $t = 0$, so the initial condition $u(0) = u_0$ is also satisfied.

Although some properties of the scalar exponential are shared by the matrix exponential, others are not. For example, although $e^{2Mt} = e^{Mt}e^{Mt}$, it is not true in general that $e^{(M_1+M_2)t} = e^{M_1 t}e^{M_2 t}$, see Exercise 4.10.

Example 4.9 Find e^{Mt} for the matrix

$$M = \begin{pmatrix} -1 & 1 \\ 0 & -2 \end{pmatrix},$$

(a) using the solution of Example 4.3,
(b) using diagonalisation.

(a) The general solution to the linear system of Example 4.3 is

$$\begin{pmatrix} x \\ y \end{pmatrix} = Ae^{-t} \begin{pmatrix} 1 \\ 0 \end{pmatrix} + Be^{-2t} \begin{pmatrix} 1 \\ -1 \end{pmatrix}.$$

If the initial condition is $x = x_0$, $y = y_0$ then $x_0 = A + B$ and $y_0 = -B$, so $B = -y_0$ and $A = x_0 + y_0$. So

$$x = (x_0 + y_0)e^{-t} - y_0e^{-2t}, \qquad y = y_0e^{-2t},$$

and this can be written in matrix form as

$$\begin{pmatrix} x \\ y \end{pmatrix} = \begin{pmatrix} e^{-t} & e^{-t} - e^{-2t} \\ 0 & e^{-2t} \end{pmatrix} \begin{pmatrix} x_0 \\ y_0 \end{pmatrix}.$$

Comparing this with the general solution $u = e^{Mt}u_0$, given above, the matrix here is the matrix exponential,

$$e^{Mt} = \begin{pmatrix} e^{-t} & e^{-t} - e^{-2t} \\ 0 & e^{-2t} \end{pmatrix}.$$

(b) The matrix M can be diagonalised by writing $M = PDP^{-1}$, where P is the matrix whose columns are the eigenvectors of M and D is a diagonal matrix containing the eigenvalues (in the same order as the eigenvectors in P). So in this case, where $P^{-1} = P$,

$$M = \begin{pmatrix} -1 & 1 \\ 0 & -2 \end{pmatrix} = \begin{pmatrix} 1 & 1 \\ 0 & -1 \end{pmatrix} \begin{pmatrix} -1 & 0 \\ 0 & -2 \end{pmatrix} \begin{pmatrix} 1 & 1 \\ 0 & -1 \end{pmatrix}.$$

Now $M^2 = PDP^{-1}PDP^{-1} = PD^2P^{-1}$, and similarly $M^n = PD^nP^{-1}$, so

$$e^{Mt} = Pe^{Dt}P^{-1} = \begin{pmatrix} 1 & 1 \\ 0 & -1 \end{pmatrix} \begin{pmatrix} e^{-t} & 0 \\ 0 & e^{-2t} \end{pmatrix} \begin{pmatrix} 1 & 1 \\ 0 & -1 \end{pmatrix} = \begin{pmatrix} e^{-t} & e^{-t} - e^{-2t} \\ 0 & e^{-2t} \end{pmatrix}.$$

Although these two methods appear at first sight to be different, they are in fact very closely related. In part (a), solving for A and B in terms of x_0 and y_0 is equivalent to finding the matrix P^{-1}. A third way to calculate e^{Mt} is to use the definition (4.14) directly, see Exercise 4.9.

Key Points from Chap. 4

- A second-order differential equation is equivalent to two coupled first-order differential equations for two dependent variables $x(t)$ and $y(t)$.
- A coupled linear system can be written in matrix form, and its behaviour depends on the eigenvalues of that matrix.
- The qualitative behaviour of two coupled linear equations can be sketched in a *phase plane*, with axes corresponding to x and y.
- The *nullclines*, lines on which $\dot{x} = 0$ or $\dot{y} = 0$, are helpful in plotting the phase plane.
- If the eigenvalues are real, there are solution trajectories in the form of straight lines in the phase plane, along the eigenvectors.
- The qualitative behaviour can be classified into several different types, depending on the eigenvalues.
- If both eigenvalues are positive, the system is described as an *unstable node*. If they are both negative, it is a *stable node*.
- If there is one positive eigenvalue and one negative one, the system is called a *saddle*.
- If the eigenvalues are a complex conjugate pair, the system is a *spiral*, stable if the real part is negative and unstable if it is positive.
- Special cases include a *degenerate node*, in which there is only one eigenvalue, and a *centre*, when the eigenvalues are purely imaginary.
- A system is *Lyapunov stable* if for initial conditions close to the origin, solutions remain close to the origin for all time. *Asymptotic stability* is a stronger condition requiring that solutions also tend to the origin as $t \to \infty$.
- Stable nodes and stable spirals are asymptotically stable. Centres are Lyapunov stable but not asymptotically stable.
- The solution of a linear system of differential equations can be written down directly in terms of the *exponential of the matrix* of the system, which is defined as a series, in a similar way to the exponential of a scalar.

Exercises

4.1 Write the second-order differential equation

$$\ddot{x} - 2\dot{x} + 2x = 0$$

as two coupled first-order equations,

(a) by using the method given at the start of this chapter;
(b) by defining $y = x - \dot{x}$.
(c) Find the characteristic equation for the eigenvalues of the matrices for the linear systems in (a) and (b) and comment briefly on the result.
(d) What type of system is this, according to the classification of Sect. 4.3?

4.2 Using eigenvalues and eigenvectors as in Example 4.1, find the solution of the system

$$\dot{x} = x - 6y,$$
$$\dot{y} = 2x - 6y,$$

with the initial conditions $x(0) = 1$, $y(0) = 0$. Classify the type of the system.

4.3 Classify each of the following systems. If the eigenvalues are real, find the eigenvectors. If the eigenvalues are complex, investigate whether the rotation is clockwise or anticlockwise.

(a) $\dot{x} = x, \ \dot{y} = -y.$
(b) $\dot{x} = 2y, \ \dot{y} = -3x.$
(c) $\dot{x} = -3x + y, \ \dot{y} = -2x.$
(d) $\dot{x} = 5x + 4y, \ \dot{y} = -x + y.$
(e) $\dot{x} = 4x - 9y, \ \dot{y} = x + 8y.$

4.4 Consider the general linear system

$$\begin{pmatrix} \dot{x} \\ \dot{y} \end{pmatrix} = \begin{pmatrix} a & b \\ c & d \end{pmatrix} \begin{pmatrix} x \\ y \end{pmatrix}.$$

(a) If it is known that the constants a, b, c and d are all positive, which of the types of system described in this chapter are possible?
(b) If the phase portrait is a spiral, show that b and c must have opposite signs, and deduce a test, using the signs of b and c, to determine whether the spiral has a clockwise or an anticlockwise rotation.
(c) If the matrix is symmetric, what types of phase portrait are possible?

4.5 For the system

$$\dot{x} = 4x - 3y, \qquad \dot{y} = 10x - 7y,$$

show that all solutions tend to the origin. Show that trajectories starting from the point (x_0, y_0) approach the origin along a particular line in the phase space as $t \to \infty$, unless x_0 and y_0 are related in a certain way.

4.6 Classify the linear system

$$\dot{x} = -x + 2y,$$
$$\dot{y} = -2x + 2y.$$

Find the nullclines and sketch them, and hence sketch the phase portrait.

4.7 (a) Show that the system

$$\dot{x} = x + y,$$
$$\dot{y} = y,$$

represents a degenerate node.

(b) By solving the y equation directly, then solving the x equation using the integrating factor method, find $x(t)$ and $y(t)$.

(c) By finding dy/dx on a solution trajectory, show that trajectories become closely aligned with the direction of the single eigenvector as $t \to \infty$ and as $t \to -\infty$.

4.8 Write down a matrix that has zero trace and zero determinant, with all of the matrix entries non-zero. This special case is not mentioned in Sect. 4.3.4. By dividing $\dot{y}$ by $\dot{x}$ to find dy/dx on a trajectory, or otherwise, sketch the phase portrait.

4.9 For the matrix M given in Example 4.9, find M^2 and M^3. Make a guess for a formula for M^n and show that your guess is correct using mathematical induction. Hence obtain e^{Mt} directly from the definition (4.14).

4.10 Consider the matrices

$$M_1 = \begin{pmatrix} 0 & 0 \\ 1 & 0 \end{pmatrix}, \quad M_2 = \begin{pmatrix} 0 & 1 \\ -1 & 0 \end{pmatrix}.$$

(a) Evaluate $e^{M_1 t}$ and $e^{M_2 t}$, using the definition (4.14).
(b) Find $e^{M_1 t} e^{M_2 t}$. Is it equal to $e^{(M_1 + M_2)t}$?

Chapter 5
Second-Order Nonlinear Systems

This chapter describes qualitative methods for systems of two coupled first-order nonlinear differential equations. In almost all cases, these systems cannot be solved analytically. As in Chap. 3, the aim is to describe the behaviour of the system and sketch solutions. For a second-order system, solutions are sketched in a two-dimensional phase plane. This process involves finding the fixed points of the system, linearising the system near each fixed point, using the methods of Chap. 4 to classify each fixed point, and then sketching the phase plane. There are several useful techniques that help with the construction of the phase plane. Applications, described through worked examples, include models for competition between different species, the motion of a pendulum, and the growth and eventual decline of an epidemic.

5.1 Introduction and Applications

This chapter is concerned with systems of two coupled, autonomous, nonlinear differential equations of the form

$$\dot{x} = f(x, y), \tag{5.1}$$

$$\dot{y} = g(x, y). \tag{5.2}$$

Here, x and y are dependent variables, functions of the independent variable t. The system (5.1) and (5.2) is *autonomous*, because t does not appear explicitly. It is assumed that there is at least one nonlinear term in the system (the linear case was covered in Chap. 4). The two first-order equations (5.1) and (5.2) are sometimes derived from a single second-order differential equation. For this reason, they are often described as a *second-order nonlinear system*. It is always easy to write a

© The Author(s), under exclusive license to Springer Nature Switzerland AG 2025
P. C. Matthews, *Differential Equations, Bifurcations and Chaos*,
Springer Undergraduate Mathematics Series,
https://doi.org/10.1007/978-3-031-99543-9_5

second-order differential equation as two first-order ones (see Sect. 5.1.2 below for an example), but the reverse process may be difficult or impossible because of the nonlinear terms. Except in a few very special cases, there is no method to find an analytical solution of the system (5.1) and (5.2).

However, we can use a qualitative approach, as we did in Chap. 3 for first-order equations. The aim of this chapter is to develop methods for plotting the phase plane for (5.1) and (5.2). Recall from Chap. 4 that the phase plane is a plot in the x, y plane showing solution trajectories of the system. Having plotted the phase plane, the behaviour of the nonlinear system can be described qualitatively, for any choice of initial condition. To provide some motivation, we will start with three applications that show how (5.1) and (5.2) can arise in mathematical modelling.

5.1.1 *Population Models*

Example 3.4 investigated a simple first-order differential equation, the logistic model, that is often used to model the growth and subsequent levelling out of a population of individuals of a single species. Similar models can describe the interaction between several different species. In the case of two species, this leads to second-order systems of the form (5.1) and (5.2). These are sometimes called *Lotka–Volterra models*, after the two mathematicians who developed them independently in the 1920s.

Suppose that two species, x and y, are present in the same environment, and assume that if one species were absent, or died out, the other one would be governed by a logistic equation of the form (3.9). If the two species did not interact with each other, we would just have two uncoupled logistic equations. But if the two species compete with each other for some resource, for example, rabbits and sheep competing for grass to eat, then there are some interaction terms in the equations. The presence of rabbits eating the grass may reduce the growth rate of the sheep population, and vice versa. This leads to models of the form

$$\dot{x} = r_1 x - b_1 x^2 - c_1 xy, \tag{5.3}$$

$$\dot{y} = r_2 y - b_2 y^2 - c_2 xy, \tag{5.4}$$

where all of the six constants in the model are positive. If $y = 0$ at any time, then $\dot{y} = 0$, so $y(t) = 0$ for all t, while $x(t)$ obeys a logistic equation, and vice versa. This system exhibits several possible types of outcome for the competition between the species, which will be explored in Example 5.8. A slightly different model arises when one species is a predator and the other is its prey (see Exercise 5.9).

5.1.2 *Newtonian Mechanics and the Nonlinear Pendulum*

Newtonian mechanics provides a rich source of second-order differential equations, which may be linear or nonlinear. Newton's second law of motion,

$$\text{force} = \text{mass} \times \text{acceleration}, \tag{5.5}$$

is, in effect, a second-order differential equation. Suppose that an object of mass M moving in one space dimension, x, is acted on by a force that is a function of its position and its velocity, $F(x, \dot{x})$. Then (5.5) becomes

$$M\ddot{x} = F(x, \dot{x}). \tag{5.6}$$

The motion of almost anything—a car, a planet, a football, even a weather system, is governed by (5.6) or extensions of it.

A specific example is the motion of a pendulum. Suppose that a pendulum bob of mass M is suspended from a pivot point by a light rod of length L, and let x be the angle that the pendulum makes with the downward vertical direction. Then, since x is an angular variable, the velocity of the pendulum bob is $L\dot{x}$ and its acceleration in the angular direction is $L\ddot{x}$. Suppose that the friction associated with the pivot mechanism, combined with air resistance, produces a drag force opposing the motion, proportional to the pendulum velocity, $-\alpha\dot{x}$, for some positive constant α. The downward force due to gravity on the pendulum bob is Mg, where g is the acceleration due to gravity, and the component of this acting in the angular direction is $-Mg\sin x$. Then (5.6) gives

$$ML\ddot{x} = -\alpha\dot{x} - Mg\sin x. \tag{5.7}$$

The second-order equation (5.7) can be written as a system of two first-order equations in the form of (5.1) and (5.2) simply by defining $y = \dot{x}$. Then we have

$$\dot{x} = y, \tag{5.8}$$

$$\dot{y} = -\beta y - \frac{g}{L}\sin x, \tag{5.9}$$

where $\beta = \alpha/ML$.

The system (5.8), (5.9) cannot be solved analytically, because of the nonlinear term $\sin x$.[1] A commonly used simplifying assumption is that the oscillations of the pendulum are small, so the approximation $\sin x \approx x$ can be used, resulting in a linear system that can be solved using the methods of Sect. 2.4 or Sect. 4.1. In the simplest case where this linearisation is assumed to be valid and there is no

---

[1] There are claims that it can be solved analytically, but only by effectively defining a new function to be what is needed to solve the equations, which is cheating.

friction ($\beta = 0$), the solution if the pendulum starts at $x = 0$ is $x = A \sin \sqrt{g/L}t$, which leads to the result that the oscillation period of the pendulum is $2\pi \sqrt{L/g}$, independent of the (small) amplitude A. The full nonlinear system (5.8), (5.9) will be investigated in Example 5.9.

5.1.3 Epidemic Modelling

Modelling the growth of infection in an epidemic is another of the many applications of differential equations. We will consider a simple model proposed by Kermack and McKendrick, published by the Royal Society in 1927. Their model has been extended in many ways and widely used, for example to model the COVID epidemic that spread throughout the world during 2020. It is known as the SIR model, since the population is divided into the proportion that are susceptible to a disease, S, the proportion that are infected, I, and the proportion that have recovered, R. It is assumed that those who are susceptible to the disease become infected when they interact with those who are infected at a rate a. Since infection requires both susceptible and infected individuals, this infection rate is modelled by the nonlinear term aSI. The infection rate increases the number of infected individuals and decreases the number of susceptibles. Those who are infected with the disease are assumed to recover at some rate b, leading to a term bI appearing with a plus sign in the equation for R and a minus sign in the equation for I. Putting these assumptions together gives the three equations

$$\dot{S} = -aSI, \tag{5.10}$$

$$\dot{I} = aSI - bI, \tag{5.11}$$

$$\dot{R} = bI. \tag{5.12}$$

Adding together all three equations gives

$$\frac{d}{dt}(S + I + R) = 0, \tag{5.13}$$

which says that the total number of individuals in the population is constant. This is due to the fact that there are no births or deaths included in this model. Since S, I and R are proportions of the total population, $S + I + R = 1$, and we can use this fact to eliminate any one of the variables and reduce the order of the system from three to two. Choosing to eliminate $R = 1 - S - I$ is simplest; in fact we can just consider the first two equations (5.10), (5.11), because R does not appear in either of these equations. The behaviour of this system will be investigated in Example 5.10.

5.2   Fixed Points

A fixed point of a second-order system is defined in a similar way to the case of a first-order differential equation discussed in Sect. 3.2. The point (x_0, y_0) is a *fixed point* or *equilibrium point* of (5.1) and (5.2) if and only if

$$f(x_0, y_0) = 0 \quad \text{and} \quad g(x_0, y_0) = 0.$$

As in the first-order case, fixed points are solutions of the system in which the dependent variables do not change.

Finding fixed points involves solving two simultaneous equations. But since the equations are generally nonlinear, there may be several solutions, and care is needed to make sure all of the possible fixed points have been found. A useful approach is to start by factorising the functions $f(x, y)$ and $g(x, y)$ as far as possible, and then consider the various possibilities systematically.

Example 5.1 Find all the fixed points of the Lotka–Volterra system

$$\dot{x} = 3x - x^2 - xy = x(3 - x - y),$$
$$\dot{y} = 4y - 2y^2 - xy = y(4 - 2y - x).$$

The first step is to factorise, as shown above. Since both $f(x, y)$ and $g(x, y)$ factorise into two terms, there are two ways to make each of them zero. Hence, in the search for fixed points, there are four possibilities to consider:

(a) $x = 0$ and $y = 0$,
(b) $x = 0$ and $4 - 2y - x = 0$, so $y = 2$,
(c) $3 - x - y = 0$ and $y = 0$, so $x = 3$,
(d) $3 - x - y = 0$ and $4 - 2y - x = 0$, leading to $x = 2$, $y = 1$.

So there are four fixed points: $(x_0, y_0) = (0, 0)$, $(0, 2)$, $(3, 0)$, $(2, 1)$. In the context of the species competition model of Sect. 5.1.1, the first represents extinction of both species, the second and third mean that only one species is present, and the last one corresponds to a steady state with both species coexisting.

5.3   Linearisation and Stability of Fixed Points

Recall from Definitions 4.1 and 4.2 that a fixed point is *Lyapunov stable* if trajectories that start close to the fixed point remain close to it, and *asymptotically stable* if it is Lyapunov stable and trajectories that start sufficiently close to the fixed point approach it as $t \to \infty$. A convention that we will use is that if a fixed point is described simply as stable, we mean Lyapunov stable. If a fixed point is not Lyapunov stable, it is unstable.

The investigation of the stability of fixed points of (5.1) and (5.2) follows the same approach as in Sect. 3.2.1 for first-order equations, but here the calculation is more involved and there are more special cases to consider. Fortunately, most of the work has already been done in Chap. 4.

Suppose that (x_0, y_0) is a fixed point of (5.1) and (5.2). Now consider a point near the fixed point, $x(t) = x_0 + u(t)$, $y(t) = y_0 + v(t)$, where u and v are both very small. Then $\dot{x} = \dot{u}$ and $\dot{y} = \dot{v}$. The functions f and g, if they are sufficiently smooth, can be expanded as a Taylor series in the two variables u and v, so

$$\dot{u} = f(x_0 + u, y_0 + v) = f(x_0, y_0) + u\left.\frac{\partial f}{\partial x}\right|_{x_0, y_0} + v\left.\frac{\partial f}{\partial y}\right|_{x_0, y_0} + O(2),$$

$$\dot{v} = g(x_0 + u, y_0 + v) = g(x_0, y_0) + u\left.\frac{\partial g}{\partial x}\right|_{x_0, y_0} + v\left.\frac{\partial g}{\partial y}\right|_{x_0, y_0} + O(2),$$

where the notation involving the vertical line is used to indicate that the partial derivatives of f and g are evaluated at the fixed point, and the $O(2)$ notation represents the remaining terms in the Taylor series, which are all proportional to products or higher powers of the two small quantities u and v.[2] Since (x_0, y_0) is a fixed point, the terms $f(x_0, y_0)$ and $g(x_0, y_0)$ are both zero. The *linearisation* of the system near the fixed point is obtained by dropping the higher terms in the Taylor series, leaving the linear system

$$\dot{u} = u\left.\frac{\partial f}{\partial x}\right|_{x_0, y_0} + v\left.\frac{\partial f}{\partial y}\right|_{x_0, y_0},$$

$$\dot{v} = u\left.\frac{\partial g}{\partial x}\right|_{x_0, y_0} + v\left.\frac{\partial g}{\partial y}\right|_{x_0, y_0}.$$

This can be written neatly in matrix form,

$$\begin{pmatrix} \dot{u} \\ \dot{v} \end{pmatrix} = J \begin{pmatrix} u \\ v \end{pmatrix} \qquad \text{where} \qquad J = \begin{pmatrix} \frac{\partial f}{\partial x} & \frac{\partial f}{\partial y} \\ \frac{\partial g}{\partial x} & \frac{\partial g}{\partial y} \end{pmatrix}. \tag{5.14}$$

The matrix J in (5.14) is known as the *Jacobian matrix*, or simply the *Jacobian* of the system (5.1) and (5.2).[3] To save writing, the vertical line notation has been dropped, but still applies, so the partial derivatives are all evaluated at the fixed point. This means that the entries in the Jacobian matrix are simply known numbers.

---

[2] See Appendix Sect. A.1 for the O notation and Sect. A.5 for a summary of partial derivatives and Taylor series.

[3] The Jacobian is pronounced with the stress on the second syllable. It is named after the nineteenth century German mathematician Carl Jacobi (stress on the first syllable).

Table 5.1 Classification of fixed points

Type	Eigenvalues	Trace and determinant
Unstable node	Both real and positive	$T^2 > 4D, D > 0, T > 0$
Stable node	Both real and negative	$T^2 > 4D, D > 0, T < 0$
Saddle	Both real, opposite signs	$D < 0$
Unstable spiral	Complex, positive real part	$T^2 < 4D, T > 0$
Stable spiral	Complex, negative real part	$T^2 < 4D, T < 0$
Degenerate node	Only one, one eigenvector	$T^2 = 4D$
Star node	Only one, many eigenvectors	$T^2 = 4D$
Centre	Imaginary eigenvalues	$T = 0, D > 0$
Line of fixed points	One eigenvalue is zero	$D = 0$

The linearised system (5.14) is precisely the system (4.2) studied in detail in Chap. 4. Therefore, each fixed point of the nonlinear system (5.1) and (5.2) can be classified into one of the types described in Sect. 4.3, after finding the Jacobian matrix J and evaluating it at the fixed point. Recall that this classification is determined by the eigenvalues of J, or alternatively by the trace T and determinant D of J. However, it is important to remember that for the nonlinear system, this description is only valid very close to the fixed point, because (5.14) comes from only keeping the linear terms in the Taylor series expansion around the fixed point.

The various possible cases are listed in Table 5.1, summarising the results found in Chap. 4. There are five generic cases, given in the first five rows of the table, where the conditions on T and D are inequalities. The remaining four rows describe special cases, where an equality is required. These special cases occur fairly frequently, since certain problems have some special features which lead directly to these equality conditions, as we shall see in the examples later in this chapter. These cases can be subdivided further; for example, the degenerate node, the star node and the line of fixed points all have stable and unstable versions.

The special case of the centre has eigenvalues that are purely imaginary, with zero real part, and therefore is on the borderline between stability and instability. A centre is Lyapunov stable, since solutions that start near the fixed point remain near it, but not asymptotically stable, because solutions do not tend towards the fixed point. In this special case, the linear system may not correctly describe the nonlinear system near the fixed point. The following theorem makes this more precise.

Theorem 5.1 (Hartman–Grobman) *If no eigenvalue of the Jacobian evaluated at a fixed point has zero real part, then there is a neighbourhood of the fixed point in which the linearised system and the nonlinear system are topologically equivalent.*

Here, *topologically equivalent* means that there is an invertible function that maps one to the other, sufficiently near the fixed point. More informally, the theorem says that the linearisation describes the behaviour near the fixed point qualitatively correctly as long as there is no eigenvalue with zero real part. A fixed point with this property is sometimes known as a *hyperbolic* fixed point.

The theorem won't be proved here, but the reasoning is that the linearisation (5.14) is a valid approximation provided that the small nonlinear terms that have been dropped do not qualitatively alter the behaviour. In the case of, for example, exponential decay to the fixed point this is true, as the neglected terms only make a slight change to the rate at which a trajectory approaches the fixed point. But for a centre, trajectories form closed orbits around the fixed point, and a small change due to nonlinear terms may cause the orbits to fail to close exactly, leading to a qualitatively different picture.

For the types of fixed point in the first seven rows of Table 5.1, the Hartman–Grobman theorem tells us that the nonlinear system behaves in a similar way to the linear system near the fixed point. But for the last two rows, the centre and the line of fixed points, there is an eigenvalue with zero real part. Then the linear and nonlinear systems may be qualitatively different, as shown in the following example.

Example 5.2 Sketch the phase portrait of the system

$$\dot{x} = y - x^2, \tag{5.15}$$

$$\dot{y} = -y, \tag{5.16}$$

first for the linearised system, and then for the nonlinear system.

The first step is to find the fixed points. From (5.16), $y = 0$ at a fixed point, and then (5.15) shows that a fixed point must have $x = 0$, so $(0, 0)$ is the only fixed point.

When the fixed point being considered is the origin, it is not necessary to find the Jacobian to investigate its stability. The system can be linearised simply by keeping the linear terms but discarding the nonlinear terms. Hence the linearised form of (5.15), (5.16) is $\dot{x} = y$, $\dot{y} = -y$. We can see that this linear system is qualitatively different from the nonlinear one, since in the linear system any point where $y = 0$ is a fixed point. So the linear system has an infinite line of fixed points, but the nonlinear system has only one fixed point. In matrix form, the linearised system is

$$\begin{pmatrix} \dot{x} \\ \dot{y} \end{pmatrix} = \begin{pmatrix} 0 & 1 \\ 0 & -1 \end{pmatrix} \begin{pmatrix} x \\ y \end{pmatrix}. \tag{5.17}$$

One of the eigenvalues of the matrix is zero, as expected. If this was not the case, this example would contradict Theorem 5.1. The corresponding eigenvector is $(1, 0)^T$, pointing along the x axis. The other eigenvalue is -1, with eigenvector $(1, -1)^T$. The last line of Table 5.1 applies and the linear system has a line of fixed points, along the x axis in this example. Away from the x axis, trajectories approach the x axis along the direction of the eigenvector $(1, -1)^T$. Since $dy/dx = \dot{y}/\dot{x} = -1$ in the linearised system, the trajectories are all of the form $y = -x + c$ for some constant c. The linear system has the phase plane picture shown in Fig. 5.1a. The fixed points on $y = 0$ are Lyapunov stable but not asymptotically stable, because for any of the fixed points, a trajectory that starts nearby will in most cases approach a different fixed point on the axis.

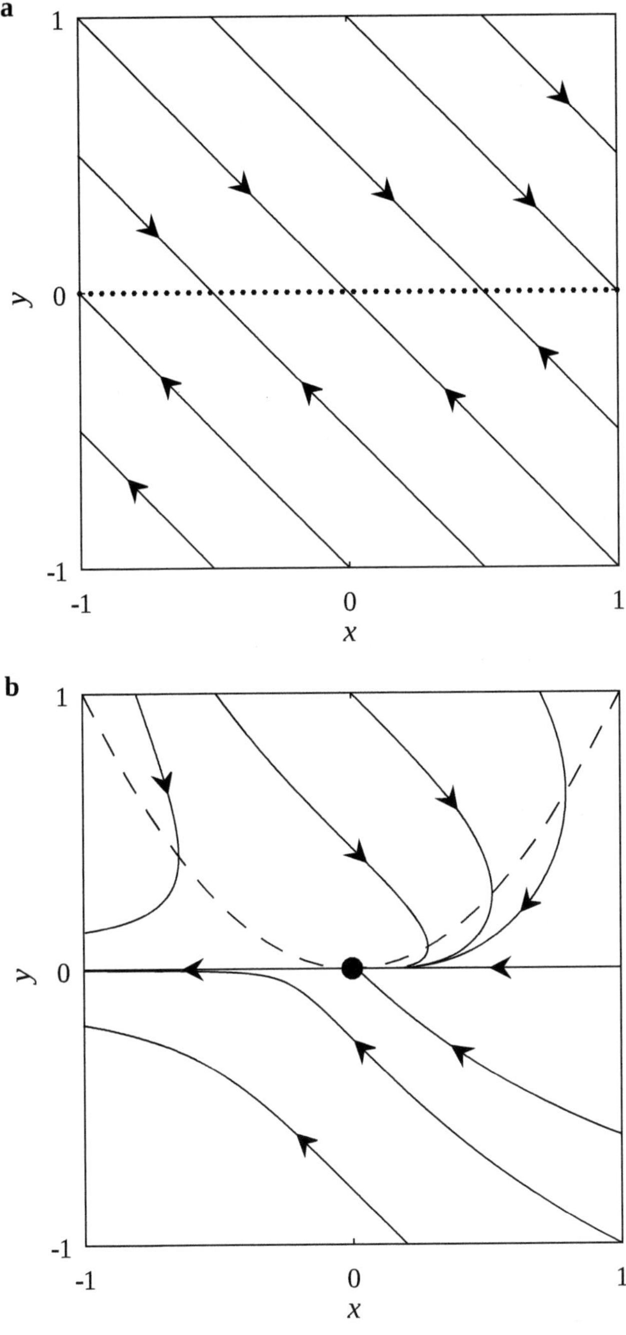

Fig. 5.1 Phase planes for Example 5.2. (**a**) In the linearised system, there is a line of fixed points (dotted line) along the x axis. (**b**) In the nonlinear system, the only fixed point is at the origin. The dashed line is the x-nullcline

In the nonlinear system (5.15), (5.16), consider a trajectory starting on the x axis, $y = 0$. Then $\dot{y} = 0$, so the trajectory remains on $y = 0$, and $\dot{x} = -x^2$. If $x > 0$, x decreases and the trajectory approaches the fixed point at the origin. But if $x < 0$, x decreases, moving away from the fixed point, as shown in Fig. 5.1b, so the origin is not Lyapunov stable in the nonlinear system, confirming that the linear and nonlinear systems are qualitatively different.

Note that the Hartman–Grobman theorem is not "if and only if". If there are no eigenvalues with zero real part, the nonlinear and linear systems are qualitatively the same near the fixed point. But if there is an eigenvalue with zero real part, the two systems may be qualitatively different, as in the example above, or they may be qualitatively the same. For example, if (5.15) is changed to $\dot{x} = y - y^2$, then there is a line of fixed points in both the linear and nonlinear systems.

Closely related to the question of the qualitative similarity between the nonlinear and linearised system is the idea of structural stability. A system is said to be *structurally stable* if a very small change to the form of the equations does not change the qualitative picture. Again, nodes, saddles and spirals are structurally stable, but centres and lines of fixed points are not, because making a small change to the system can significantly change the phase portrait picture.

The aim of this chapter is to sketch the phase plane of the nonlinear system (5.1) and (5.2). A key step in this process is to find and classify the fixed points of the system, as discussed above. However, there are several other useful techniques that can help us to sketch the phase plane. Three of these are described in the following sections.

5.4 Nullclines

The idea of a nullcline has already been introduced in Sect. 4.2. In that section, on linear systems, the nullclines were always straight lines. But for nonlinear systems, the nullclines are generally curves. The *x-nullcline* for the system (5.1) and (5.2) is the curve in the x, y plane on which $\dot{x} = 0$, in other words, the curve given by $f(x, y) = 0$. On this curve, trajectories point vertically in the x, y phase plane plot, either upwards or downwards, depending on the sign of $\dot{y}$. For example, in (5.15) the x-nullcline is $y = x^2$, shown as a dashed line in Fig. 5.1b, which is helpful for sketching the phase portrait. Similarly, the *y-nullcline* is the curve on which $\dot{y} = g(x, y) = 0$, and on this curve trajectories are horizontal. Where the x- and y-nullclines intersect, $\dot{x} = 0$ and $\dot{y} = 0$, so this point is a fixed point. In general, each nullcline may consist of more than one curve, as shown by the following example.

Example 5.3 Find the nullclines of the system

$$\dot{x} = \frac{2}{1 + y^2} - x, \tag{5.18}$$

$$\dot{y} = y(x - y^2). \tag{5.19}$$

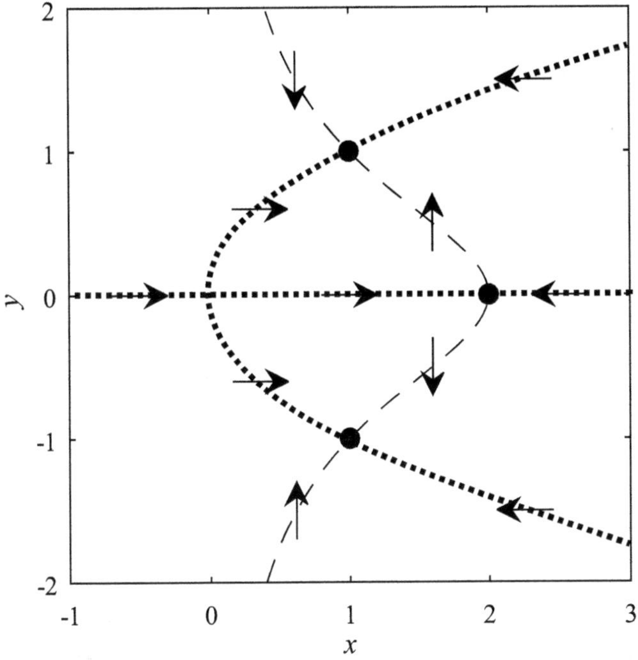

Fig. 5.2 Nullclines for (5.18), (5.19). The dashed line is the x-nullcline and the dotted lines are the two parts of the y-nullcline. Arrows indicate the direction of trajectories on each nullcline, and the large dots are the three fixed points

Sketch the nullclines, showing the direction of trajectories as they cross each nullcline. Identify the fixed points, and use the diagram of the nullclines to guess the type of each one.

The x-nullcline is the curve $x = 2/(1+y^2)$. The y-nullcline is where $y(x-y^2) = 0$, so it consists of two parts, the line $y = 0$ and the curve $x = y^2$. The nullclines are shown in Fig. 5.2.

Consider first the x-nullcline. On this curve,

$$\dot{y} = y\left(\frac{2}{1+y^2} - y^2\right) = \frac{y(2 - y^2 - y^4)}{1+y^2} = \frac{y(2 + y^2)(1 - y^2)}{1+y^2},$$

so $\dot{y} > 0$ for $y < -1$ and for $0 < y < 1$, but $\dot{y} < 0$ for $y > 1$ and for $-1 < y < 0$, as shown in the figure. On the first part of the y-nullcline, where $y = 0$, $\dot{x} = 2 - x$, so $\dot{x} > 0$ for $x < 2$ and $\dot{x} < 0$ for $x > 2$. For the other part of the y-nullcline, we can find the direction of trajectories by substituting $x = y^2$ into (5.18), but there is a shortcut. We already know that $\dot{x} > 0$ on the x axis for $x < 2$, and also that $\dot{x} = 0$ only on the nullcline $x = 2/(1 + y^2)$. Therefore, for all points on $x = y^2$ that are to the left of the x-nullcline curve, $\dot{x}$ is positive, but to the right of that curve it is negative. Hence the direction of trajectories on the nullclines is as shown in Fig. 5.2.

The fixed points are where the nullclines cross, so there are three: $(2, 0)$, $(1, 1)$ and $(1, -1)$. The origin $(0, 0)$ is not a fixed point. Although there appears to be a crossing of nullclines at the origin, this is a crossing of the two parts of the y-nullcline, not a crossing of the x- and y-nullclines as required for a fixed point.

Finally, looking at the nullcline diagram gives clues about the type of each fixed point. Near $(2, 0)$, trajectories approach the fixed point along the x axis, but point away from it in the y direction. This suggests that this is a saddle point. Near the fixed points at $(1, \pm 1)$, trajectories appear to be spiralling in towards the fixed point, suggesting that both of these are stable spirals. This can be checked, see Exercise 5.2.

5.5   Hamiltonian Systems

Hamiltonian systems are a special case of the general second-order system (5.1) and (5.2). In physical systems they often arise when some quantity, such as the total energy of the system, is conserved. If a system is Hamiltonian, there are restrictions on the types of trajectories and fixed points that may occur, and these restrictions simplify the process of sketching the phase plane.

Definition 5.1 A second-order system is *Hamiltonian* if it can be written as

$$\dot{x} = \frac{\partial H}{\partial y}, \qquad \dot{y} = -\frac{\partial H}{\partial x}, \tag{5.20}$$

for some differentiable function $H(x, y)$ called the *Hamiltonian function* of the system.

If a system has a Hamiltonian function, there are several consequences.

- $H(x, y)$ is a conserved quantity. To show this, consider

$$\frac{dH}{dt} = \frac{\partial H}{\partial x}\dot{x} + \frac{\partial H}{\partial y}\dot{y} = \frac{\partial H}{\partial x}\frac{\partial H}{\partial y} - \frac{\partial H}{\partial y}\frac{\partial H}{\partial x} = 0. \tag{5.21}$$

  Since $H(x, y)$ does not change with time, the trajectories of (5.20) are just the contours, or level curves, of H. The converse is not necessarily true: if a system has a conserved quantity, it does not always have Hamiltonian form.
- The flow vector $\mathbf{u} = (u, v) = (\dot{x}, \dot{y})$ is a solenoidal vector field, that is, it has zero divergence. The divergence is

$$\nabla \cdot \mathbf{u} = \frac{\partial u}{\partial x} + \frac{\partial v}{\partial y} = \frac{\partial^2 H}{\partial x \partial y} - \frac{\partial^2 H}{\partial y \partial x} = 0, \tag{5.22}$$

  using the general rule that partial derivatives commute.

- The only possible types of fixed points in a Hamiltonian system are saddles and centres. The Jacobian matrix for (5.20) is

$$
J = \begin{pmatrix} \dfrac{\partial^2 H}{\partial x \partial y} & \dfrac{\partial^2 H}{\partial y^2} \\[2mm] -\dfrac{\partial^2 H}{\partial x^2} & -\dfrac{\partial^2 H}{\partial y \partial x} \end{pmatrix}.
$$

The trace of J is zero, so from the results in Table 5.1, any fixed point must be either a saddle point, if $|J| < 0$, or a centre, if $|J| > 0$.

Example 5.4 For what parameter values are the nonlinear pendulum equations (5.8), (5.9) Hamiltonian? In this case, what is H and what is its physical significance?

The divergence of (5.8), (5.9) is $-\beta$, so if the system is Hamiltonian, $\beta = 0$. This is the case where there is no friction or damping influencing the motion. The Hamiltonian function $H(x, y)$ is found from

$$
\frac{\partial H}{\partial y} = y, \qquad -\frac{\partial H}{\partial x} = -\frac{g}{L}\sin x,
$$

and these equations can be solved to give

$$
H = \frac{y^2}{2} - \frac{g}{L}\cos x + c = \frac{\dot{x}^2}{2} - \frac{g}{L}\cos x + c,
$$

where c is an arbitrary constant. The Hamiltonian is the total energy, kinetic plus potential, of the pendulum per unit mass, and the fact that H is constant is equivalent to the statement that the energy of the pendulum is conserved if there is no damping.

Example 5.5 Given that the system

$$
\dot{x} = 8y + ax^2 y,
$$
$$
\dot{y} = -2x + 2xy^2,
$$

is Hamiltonian, what is the value of a? For this value of a, construct the Hamiltonian function $H(x, y)$ and sketch the phase plane.

If the system is Hamiltonian, its divergence, which is $2axy + 4xy$, must be zero, so $a = -2$. To find $H(x, y)$ for this value of a, solve the two equations

$$
\frac{\partial H}{\partial y} = 8y - 2x^2 y, \qquad -\frac{\partial H}{\partial x} = -2x + 2xy^2.
$$

Integrating the first equation gives $H(x, y) = 4y^2 - x^2 y^2 + h(x)$, where $h(x)$ is any function of x, known as a *function of integration*, that arises from solving the partial differential equation, analogous to the constant of integration for an ordinary differential equation. The second equation is then

$$2xy^2 - h'(x) = -2x + 2xy^2,$$

so $h(x) = x^2 + c$, for any constant c. Therefore the Hamiltonian function is

$$H(x, y) = x^2 + 4y^2 - x^2y^2 + c.$$

We can choose $c = 0$, because the value c of has no effect on (5.20).

Having found H, we can sketch the phase plane, using the facts that H is constant on a trajectory, and that any fixed points must be either centres or saddles. Clearly the system has a fixed point at the origin. Near the origin, where x and y are both small, the quartic term x^2y^2 in H is much smaller than the other two terms, so the curves $H =$ constant are approximately $x^2 + 4y^2 =$ constant, which are ellipses centred at the origin. Hence the origin must be a centre. In general it is hard to plot the curves $H =$ constant, but by writing $H = 4y^2 + (1 - y^2)x^2$, we can see that two of these curves are the straight lines $y = \pm 1$, on which $H = 4$, for any value of x. Similarly, on $x = \pm 2$, $H = 4$. These lines intersect at $(\pm 2, \pm 1)$, so these four points are fixed points, and they must be saddles. Using the fact that trajectories cannot cross except at fixed points, and using the original equations to find the direction of trajectories, we can sketch the phase plane, see Fig. 5.3.

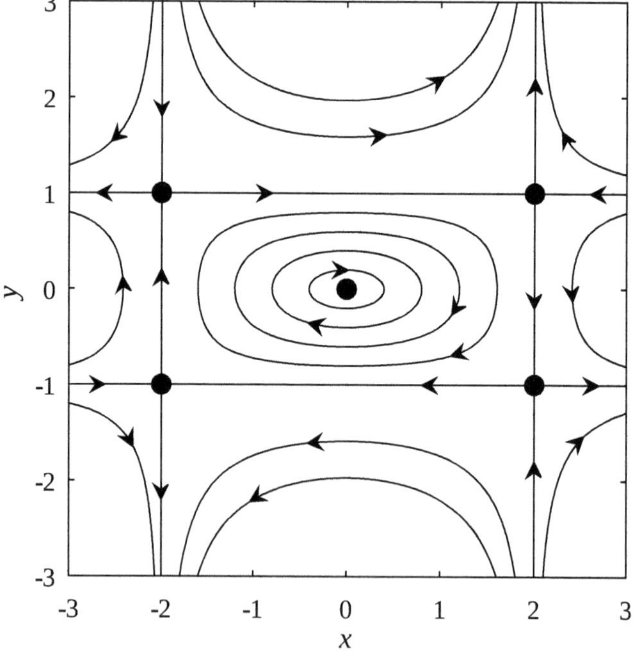

Fig. 5.3 Phase plane for the Hamiltonian system of Example 5.5. The large dots mark the fixed points, one centre and four saddles

In this example we can draw the phase plane without calculating the Jacobian matrix and finding its eigenvalues to determine the type of each fixed point. Also, we have not made use of the nullclines. Both of these methods can be used to check that the phase plane diagram in Fig. 5.3 is correct.

The special straight line trajectories in this example, that are very helpful for plotting the phase plane, can also occur in non-Hamiltonian systems. They are known as *invariant lines*, defined and discussed in the following section.

5.6   Invariant Lines and Invariant Curves

Invariant lines or invariant curves, when they exist, provide another useful tool for understanding the behaviour of the system (5.1) and (5.2) and sketching its phase plane.

Definition 5.2 The curve $h(x, y) = 0$ is an *invariant curve* for the system (5.1) and (5.2) if for all points lying on the curve $h(x, y) = 0$, $d(h(x, y))/dt = 0$. If $h(x, y) = 0$ is a straight line, it is known as an *invariant line*.

Consider a trajectory with an initial condition on an invariant curve $h(x, y) = 0$. Then since $d(h(x, y))/dt = 0$, h remains zero, so the trajectory remains on the invariant curve.

It may seem from the definition that an invariant curve is simply a trajectory of the system, but the two terms are not equivalent. An invariant curve often contains fixed points, see for example the lines $H = 4$ in Fig. 5.3, in which case the curve is not a trajectory, but contains trajectories.

Invariant lines and curves occur in three of the examples already presented. In Example 5.2, the second equation is simply $\dot{y} = -y$. Hence $y = 0$ is an invariant line, since if $y = 0$ then $\dot{y} = 0$. Therefore any trajectory that starts on the line $y = 0$ remains on that line. This line includes trajectories that start from a point where $x > 0$ and approach the origin, the origin itself, which is a fixed point, and trajectories that start with $x < 0$ and move along the $y = 0$ line approaching $x = -\infty$. Usually an invariant line includes fixed points and trajectories, but it can consist of a line of fixed points, for example in the linearised case shown in Fig. 5.1a.

In Example 5.3, the line $y = 0$ is again invariant, since $y = 0 \Rightarrow \dot{y} = 0$. This line includes the fixed point $(2, 0)$, and trajectories that approach this fixed point from the left or right. In each of these cases, the invariant line corresponds to a nullcline, but this is not generally the case.

Thirdly, in Example 5.5, the curve $H(x, y) = $ constant is invariant, because of (5.21). These invariant curves include the four invariant lines on which $H = 4$.

Example 5.6 Show that $y = x^2$ is an invariant curve for

$$\dot{x} = x + 3y + 2xy, \tag{5.23}$$

$$\dot{y} = 2y + 6xy + 4y^2. \tag{5.24}$$

First, set $h(x, y) = y - x^2$ so that the curve is $h(x, y) = 0$ as in Definition 5.2. To show that the curve is invariant, we need to show that $\dot{h} = 0$ on the curve.

$$\dot{h} = \dot{y} - 2x\dot{x} = 2y + 6xy + 4y^2 - 2x(x + 3y + 2xy) = 2y + 4y^2 - 2x^2 - 4x^2y.$$

On the curve $y = x^2$,

$$\dot{h} = 2x^2 + 4x^4 - 2x^2 - 4x^4 = 0,$$

so $y = x^2$ is an invariant curve. This example is investigated further in Exercise 5.4.

5.6.1   Reflection Symmetry and Invariant Lines

There are two types of problem in which invariant lines commonly occur. The first of these is when the system has a reflection symmetry. Suppose that (5.1) and (5.2) are symmetrical under reflection of the plane in the straight line $h(x, y) = 0$. Then the line $h(x, y) = 0$ is invariant. For example, the system in Example 5.3 has the symmetry $y \leftrightarrow -y$, because the equations are unchanged if we replace y with $-y$. This corresponds to reflection of the plane in the line $y = 0$ (see Fig. 5.2), and the line $y = 0$ is therefore invariant.

To see why this is generally the case, consider Fig. 5.4. The straight dotted line represents the line of reflection symmetry. The symmetry line does not have to be horizontal as shown in the figure, but we can always rotate the coordinate system so that it is. Suppose that a trajectory passing through a point on the symmetry line points in a particular direction, as shown by the solid arrow. Then, using the reflection symmetry, there must also be a solution trajectory corresponding to the reflected arrow, shown by the dashed arrow. But this is a contradiction, since in the system (5.1) and (5.2) the vector $(\dot{x}, \dot{y})$ is uniquely defined. Trajectories cannot cross.

This contradiction can only be escaped in two ways. Either the vector points along the line, in which case it is unchanged by the reflection, so there is no contradiction, or, the vector does not have a direction because it is the zero vector, in which case the point is a fixed point. Thus, any point on a line of symmetry is

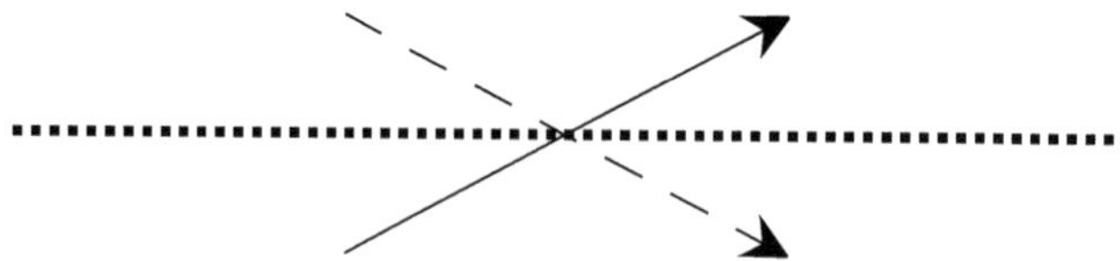

Fig. 5.4 The dotted line is a reflection symmetry of the phase plane. If the solid arrow gives the direction of a trajectory at a point on the line, then the dashed arrow must also show the direction of a trajectory

either a fixed point or a point at which a trajectory moves along the line. Hence, a line of reflection symmetry is an invariant curve.

The way in which a reflection symmetry forces the existence of an invariant line is shown algebraically in the following example.

Example 5.7 A second-order system for $x(t)$ and $y(t)$ has the symmetry $x \leftrightarrow y$. Show that the line $y = x$ is an invariant line.

Consider the general second-order system (5.1) and (5.2). Interchanging the variables x and y gives the system

$$\dot{y} = f(y, x), \qquad \dot{x} = g(y, x).$$

If the system is symmetrical under this operation, these two equations must be the same as the original two. Therefore

$$\dot{y} = f(y, x) = g(x, y) \quad \text{and} \quad \dot{x} = g(y, x) = f(x, y),$$

and so f and g are related by $f(x, y) = g(y, x)$. Now consider the function $h(x, y) = x - y$, which is zero on the line $y = x$.

$$\dot{h} = \dot{x} - \dot{y} = f(x, y) - g(x, y) = g(y, x) - g(x, y).$$

On the line $y = x$, this is $g(x, x) - g(x, x) = 0$, so $\dot{h} = 0$ on the line $h = 0$, so $y = x$ is an invariant line.

5.6.2 *Invariant Lines in Population Models*

The second situation in which invariant lines commonly occur is in population models. Suppose that the variable $x(t)$ is a measure of the population size, of, for example, oak trees, bacteria, hedgehogs or gorillas, living in an enclosed environment. Then, if at any time the population size is zero, it should remain zero for all time. Species do not come back from extinction. So $x = 0$ should be an invariant line of the system. See, for example, Eq. (5.3) in the population model of Sect. 5.1.1, where each term in the equation for $\dot{x}$ includes a factor of x, making $x = 0$ an invariant line. Since trajectories cannot cross an invariant line, if the initial condition has $x \geq 0$, x can never become negative, which is an important property that any realistic population model should have. Similarly, in the SIR epidemic model (5.10)–(5.12), the lines $S = 0$ and $I = 0$ are invariant.

5.7 Applications Revisited

This section returns to the three applications introduced in Sect. 5.1. In each case, the aim is to sketch the phase plane and hence give a qualitative description of the behaviour of the system for various choices of initial conditions. This will make use of the ideas and techniques discussed in the previous sections. Exactly which methods are most useful depends on the specific problem, but the basic guidelines are as follows. The first step is to find the fixed points and determine their type and their stability. In most cases, sketching the nullclines is also useful. If the system is Hamiltonian, sketching the phase plane is easier since the trajectories are just contour lines of the Hamiltonian function. This is likely to be the case if the system arises from a physical problem where the total energy is conserved. Invariant lines are very helpful if they exist. This is the case for most population models, and systems with a reflection symmetry.

Example 5.8 Classify the fixed points, find the nullclines, identify any invariant lines and hence sketch the phase plane for the Lotka–Volterra population model of Sect. 5.1.1 with the following parameter values:

$$\dot{x} = 3x - x^2 - xy = x(3 - x - y),$$

$$\dot{y} = 4y - 2y^2 - xy = y(4 - 2y - x).$$

In Example 5.1 the fixed points were found to be $(0, 0)$, $(0, 2)$, $(3, 0)$, $(2, 1)$. The Jacobian matrix is

$$J = \begin{pmatrix} 3 - 2x - y & -x \\ -y & 4 - 4y - x \end{pmatrix}.$$

To classify the fixed points, first evaluate J at each fixed point:

$$J_{0,0} = \begin{pmatrix} 3 & 0 \\ 0 & 4 \end{pmatrix}, \quad J_{0,2} = \begin{pmatrix} 1 & 0 \\ -2 & -4 \end{pmatrix}, \quad J_{3,0} = \begin{pmatrix} -3 & -3 \\ 0 & 1 \end{pmatrix}, \quad J_{2,1} = \begin{pmatrix} -2 & -2 \\ -1 & -2 \end{pmatrix}.$$

For $(0, 0)$ the eigenvalues are 3 and 4, so this point is an unstable node, with eigenvectors aligned with the axes. At $(0, 2)$ the eigenvalues are 1 and -4, so this is a saddle. The eigenvectors are $(5, -2)^T$ and $(0, 1)^T$. Similarly, $(3, 0)$ is a saddle with eigenvalues -3, 1 and eigenvectors $(1, 0)^T$ and $(3, -4)^T$. Finally, for $(2, 1)$, the eigenvalues are $-2 \pm \sqrt{2}$, so this point is a stable node. The eigenvectors could be found for this fixed point but this is not essential for sketching the phase plane (for saddles, it *is* always necessary to find the eigenvectors).

The x-nullcline consists of two straight lines, $x = 0$ and $x + y = 3$. Similarly the y-nullcline includes the lines $y = 0$ and $x + 2y = 4$. As a check, the intersection of these nullclines gives the fixed points already found.

Since the $\dot{x}$ equation has a factor of x, the line $x = 0$ is invariant (if $x = 0$, then $\dot{x} = 0$, so x remains at 0). Similarly, the line $y = 0$ is invariant. As discussed in Sect. 5.6, this is expected for a realistic population model.

Using the information above, we can sketch the phase plane. Since it is a population model, only the quadrant $x \geq 0$, $y \geq 0$ is of interest. Consider first solution trajectories on the invariant line $y = 0$. On this line, $\dot{x} = 3x - x^2$, so x approaches the fixed point at $(3, 0)$. This point is stable for trajectories on the invariant line, but unstable in the full system. Similarly, on the invariant line $x = 0$ solutions approach $(0, 2)$. These two points are saddles, so near each one, solutions move away from the fixed point along the unstable directions. However, this only applies near each fixed point, as it is based on linearising the system about the fixed point.

Putting all this information together, we can draw the phase plane, see Fig. 5.5. The interpretation of the phase plane is that for all initial conditions with both populations non-zero, trajectories approach the stable equilibrium in which both species coexist, $x = 2$, $y = 1$, as $t \to \infty$. For the special initial conditions in which only one species is present, that species approaches an equilibrium that is stable within the invariant line.

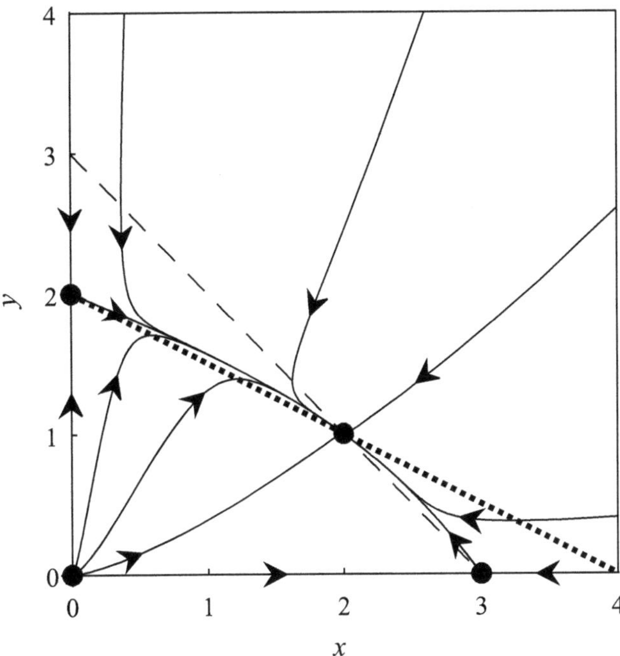

Fig. 5.5 Phase plane for Example 5.8. The dashed line is the x-nullcline and the dotted line is the y-nullcline

There are several other possible outcomes of the Lotka–Volterra model, depending on the choice of parameters. See Exercises 5.7 and 5.8 for examples.

Example 5.9 Investigate the phase plane of the nonlinear pendulum system (5.8), (5.9) in the two cases (a) $\beta = 0$, (b) $\beta > 0$.

(a) In the case $\beta = 0$ there is no damping, so the energy is conserved in the system and it should be Hamiltonian. The system is

$$\dot{x} = y, \qquad \dot{y} = -\frac{g}{L}\sin x,$$

and since $\dot{x}$ does not depend on x and $\dot{y}$ does not depend on y, the divergence condition (5.22) is automatically satisfied. Equations (5.20) can be solved to find the Hamiltonian function,

$$H(x, y) = \frac{y^2}{2} - \frac{g}{L}\cos x$$

plus an arbitrary constant. The two terms in $H(x, y)$ are proportional to the kinetic energy and the potential energy of the pendulum bob. The choice of the arbitrary constant is equivalent to the choice of what height we choose to measure the potential energy from. It can be set to zero.

Because the variable x is an angle, it is periodic with period 2π, so that the values $x = \pi$ and $x = -\pi$ are equivalent. This means that the phase "plane" in this example is really a cylinder. We can choose to consider values for x in the range $-\pi < x \leq \pi$.

The fixed points are at $(0, 0)$ and $(\pi, 0)$. The Jacobian is

$$J = \begin{pmatrix} 0 & 1 \\ -\frac{g}{L}\cos x & 0 \end{pmatrix},$$

so its trace is $T = 0$, as is always the case for Hamiltonian systems, so any fixed points are saddles or centres. For the fixed point $(0, 0)$, $D = |J| = g/L > 0$, so this point is a centre. At $(\pi, 0)$, $|J| < 0$, so this point is a saddle. Near the centre, trajectories form closed orbits around the origin, corresponding to contour lines of H. Near a saddle, it is useful to find the direction of the eigenvectors. In this case, with parameters in the Jacobian matrix, we can see that for the positive eigenvalue the eigenvector has both components of the same sign, so the vector points up and to the right or down and to the left, while for the negative eigenvalue the vector components have opposite sign. This is sufficient to sketch the phase plane, see Fig. 5.6a, which shows fixed points at $+\pi$ and $-\pi$ which are equivalent. The interpretation is as follows. Near the origin there are nested closed orbits corresponding to periodic oscillations of the pendulum, with different amplitude. There are two special trajectories that connect the two fixed points at $\pm\pi$, or equivalently the fixed point at π with itself.

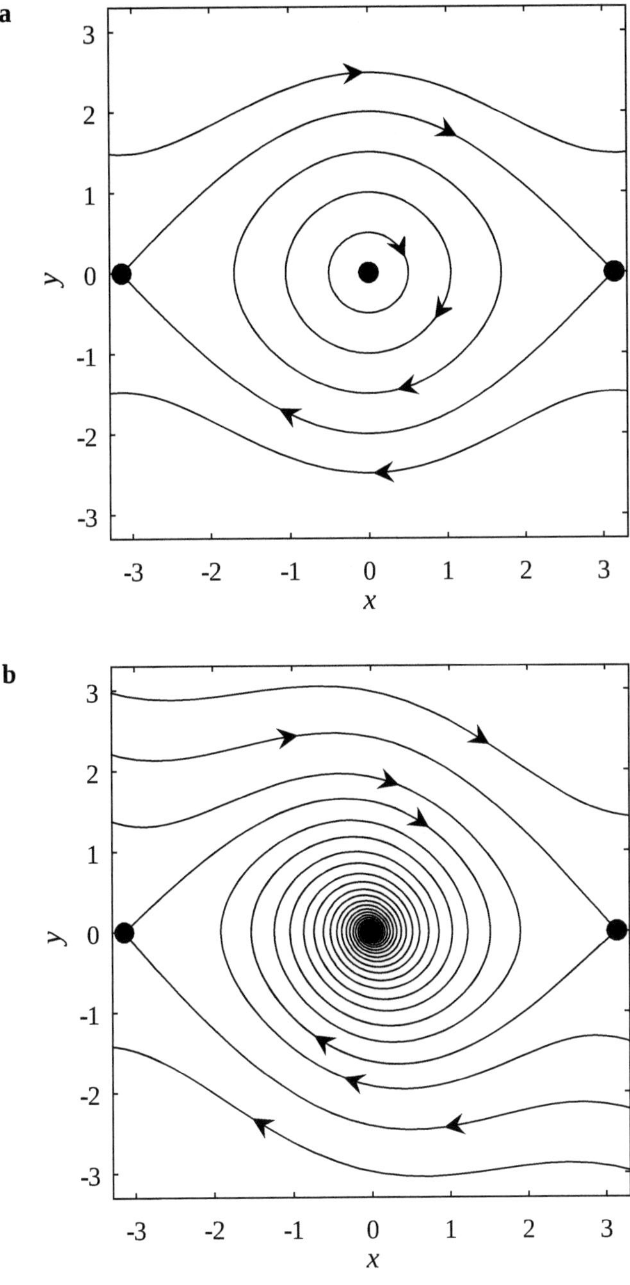

Fig. 5.6 Phase plane for the nonlinear pendulum of Example 5.9. (**a**) The undamped, Hamiltonian case, $\beta = 0$, for $g/L = 1$. The origin is a centre. (**b**) With damping, $\beta = 0.2$, $g/L = 1$. The origin is a stable spiral

Such trajectories joining two fixed points are known as *heteroclinic* trajectories. These correspond to solutions where the pendulum starts in a vertical position, and then swings down and then back to a vertical position again. The trajectories above the upper heteroclinic trajectory have $\dot{x} > 0$ for all t, so they correspond to the pendulum spinning round continuously. Similarly, there are trajectories below the lower heteroclinic solution, representing continuous rotation in the opposite direction.

(b) Now consider the case including friction, $\beta > 0$,

$$\dot{x} = y, \qquad \dot{y} = -\beta y - \frac{g}{L}\sin x.$$

The fixed points are still at $(0,0)$ and $(\pi, 0)$. The point at $(\pi.0)$ is still a saddle point, since it can easily be checked that $|J| < 0$. But the centre of the undamped case changes type for $\beta > 0$, because the trace of the Jacobian is now $-\beta < 0$, so the fixed point at $x = 0$ is no longer a centre. For small values of β it becomes a stable spiral. This fits with intuition, as the effect of friction should be to make the amplitude of oscillations gradually decrease to zero. The phase plane is shown in Fig. 5.6b for the parameters $\beta = 0.2$, $g/L = 1$. For larger values of β, the fixed point of the origin becomes a stable node, see Exercise 5.11.

Example 5.10 Find all the fixed points of the SIR model of Sect. 5.1.3,

$$\dot{S} = -aSI, \qquad \dot{I} = aSI - bI, \qquad \dot{R} = bI, \tag{5.25}$$

and determine their stability and type. Find the nullclines and any invariant lines. Hence plot the phase plane and describe the behaviour of the epidemic model.

Recall from Sect. 5.1.3 that S, I and R are the susceptible, infected and recovered proportions of the population. Although the system appears to be third-order, it is really second-order, since $S + I + R = 1$, and we can consider just the first two equations.

At a fixed point, $aSI = 0$ and $aSI - bI = 0$. These equations are both satisfied if $I = 0$, for any value of S, so the line $I = 0$ is a line of infinitely many fixed points. This makes sense, because if there are no infected individuals, everyone is either susceptible or recovered, and whatever the proportions are between these two states, nothing will change because none of the susceptibles can become infected. Since $\dot{I} = 0$ for any $I = 0$, the line $I = 0$ is also an invariant line of the system. There are no other fixed points. The Jacobian matrix is

$$J = \begin{pmatrix} -aI & -aS \\ aI & aS - b \end{pmatrix} = \begin{pmatrix} 0 & -aS \\ 0 & aS - b \end{pmatrix} \text{ on } I = 0.$$

One eigenvalue is zero, as expected for a line of fixed points, with its eigenvector pointing along the line of fixed points $I = 0$. The other eigenvalue is $aS - b$. This is positive, indicating instability, if $S > b/a$, but negative if $S < b/a$, showing stability (in the Lyapunov sense, since there is a zero eigenvalue). This means that if $S > b/a$, then introducing a small number of infected individuals will lead to exponential growth of the infection, but if $S < b/a$, so that $R > 1 - b/a$, the infection-free state is stable. So this extremely simple model captures two of the main features of epidemics such as COVID-19. When the infection first appears, I and R are small and S is large, so I grows exponentially. But if a sufficient number of people have had the infection and developed immunity, then S is small and the infection rate decreases. This is the concept known as *herd immunity*.

The S-nullcline consists of the two lines $S = 0$ and $I = 0$. Since $\dot{S} = 0$ on $S = 0$, the line $S = 0$ is an invariant line. On this line, all individuals are either infected or recovered, and I decreases exponentially so that the trajectory approaches the origin of the S, I system, which is the state where all individuals have recovered.

The I-nullcline includes the line $I = 0$ and also the line $S = b/a$. This represents the peak of an infection wave. For $S > b/a$, I increases, but when the number of uninfected people drops below this point, the number of infections starts to decrease.

The phase plane diagram is shown in Fig. 5.7. In this case the phase plane consists of a triangular region, because $S \geq 0$ and $I \geq 0$ and the constraint $R \geq 0$ means that $S + I \leq 1$. A typical epidemic starts at the bottom right of the diagram, where

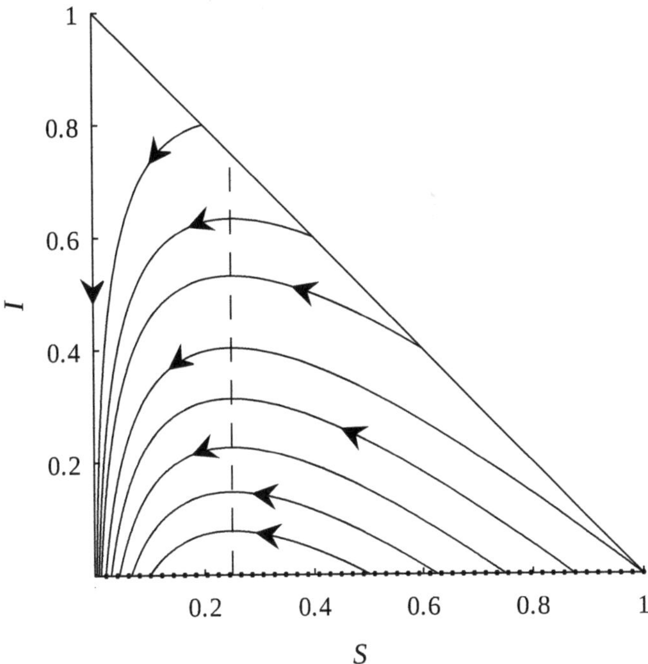

Fig. 5.7 Phase plane for the SIR model (5.25), for $a = 4$, $b = 1$. The dots along $I = 0$ mark the line of fixed points. The dashed line is the I-nullcline which marks the peak infection rate. The solid diagonal line is the boundary of the valid region, $S + I < 1$

S is high and I and R are low. I then grows until it reaches the peak when $S = b/a$, after which I decreases again.

The SIR model presented here is one of the simplest possible models for an epidemic. Many extensions to the model can be included, for example including different death rates for each group, including the effects of vaccination or social distancing policies, or including the possibility of those who have recovered losing their immunity, see Exercise 5.12.

5.8 Periodic Orbits

A *periodic orbit* or *limit cycle* is a trajectory in phase space that corresponds to a solution of (5.1) and (5.2) obeying $(x(t), y(t)) = (x(t + T), y(t + T))$ for some constant $T > 0$. If T_0 is the smallest positive value of T for which this holds, then T_0 is called the *period* of the periodic orbit.

Hamiltonian systems often have infinitely many periodic orbits, nested around each other, as in Fig. 5.6a. However, in non-Hamiltonian systems, this is usually not the case: if periodic orbits exist, they are isolated. Unfortunately, in almost all cases, it is not possible to find a periodic orbit analytically. The next section describes a simple example where this is possible, and the following sections give some conditions for which we can say either that the system cannot have any periodic orbits, or that it must have a periodic orbit.

Example 5.11 A system with a periodic orbit that can be found explicitly is

$$\dot{x} = x - y - x(x^2 + y^2), \tag{5.26}$$

$$\dot{y} = y + x - y(x^2 + y^2). \tag{5.27}$$

In this and related systems, a simplification can be made by switching to polar coordinates $r(t)$ and $\theta(t)$, related to $x(t)$ and $y(t)$ by

$$x(t) = r(t)\cos\theta(t), \qquad y(t) = r(t)\sin\theta(t).$$

Having explicitly written that x, y, r and θ are all functions of t, this dependence is now dropped to save writing. Differentiating and using the product rule and the chain rule gives

$$\dot{x} = \dot{r}\cos\theta - r\dot{\theta}\sin\theta, \qquad \dot{y} = \dot{r}\sin\theta + r\dot{\theta}\cos\theta.$$

After multiplying the first of these equations by $\cos\theta$, the second by $\sin\theta$ and adding, $\dot{\theta}$ is eliminated and we can solve for $\dot{r}$. Similarly $\dot{r}$ can be eliminated to solve for $\dot{\theta}$. The result is, for $r \neq 0$,

$$\dot{r} = \frac{x\dot{x} + y\dot{y}}{r}, \qquad \dot{\theta} = \frac{x\dot{y} - y\dot{x}}{r^2}. \tag{5.28}$$

If $\dot{x}$ and $\dot{y}$ are given by (5.26), (5.27), then

$$\dot{r} = \frac{x\left(x - y - x(x^2 + y^2)\right) + y\left(y + x - y(x^2 + y^2)\right)}{r}$$

$$= \frac{x^2 + y^2 - (x^2 + y^2)^2}{r} = r - r^3. \tag{5.29}$$

Similarly,

$$\dot{\theta} = \frac{x\left(y + x - y(x^2 + y^2)\right) - y\left(x - y - x(x^2 + y^2)\right)}{r^2}$$

$$= \frac{x^2 + y^2}{r^2} = 1. \tag{5.30}$$

The equations for $\dot{r}$ and $\dot{\theta}$ are uncoupled, so we can consider them separately. The $\dot{\theta}$ equation (5.30) states that the angle θ increases at a constant rate, so $\theta(t) = t + \text{constant}$, which corresponds to a steady anticlockwise rotation in the phase plane. The $\dot{r}$ equation (5.29) is a simple nonlinear first-order equation of the type studied in Chap. 3. It has an unstable fixed point at $r = 0$ and a stable fixed point at $r = 1$ (in polar coordinates, $r \geq 0$ so the fixed point at $r = -1$ is not relevant).

The stable fixed point at $r = 1$ of (5.29) corresponds to a circle in the plane. On this circle, $\dot{r} = 0$ but $\dot{\theta} = 1$, so there is a periodic orbit in which the trajectory follows the circle in an anticlockwise direction. The unstable fixed point at $r = 0$ is an unstable spiral in the two-dimensional system, since for solution trajectories near the origin, $\dot{r} > 0$ and $\dot{\theta} = 1$. This can easily be checked from the original system (5.26), (5.27), without using polar coordinates.

The phase plane is shown in Fig. 5.8. Trajectories that start with $r < 1$ spiral outwards, approaching the periodic orbit at $r = 1$, while those starting from $r > 1$ spiral inwards towards $r = 1$. The periodic orbit is stable, since nearby solutions approach it. In fact all solutions end up on the periodic orbit, except for the special case of an initial condition exactly at the origin. Periodic orbits can also be unstable (see Exercise 5.14).

Although this example may appear simple and contrived, it will turn out to be important in Sect. 6.2.2.

5.8.1   The Divergence Test

This section and the following three sections describe methods that can be used to show that a second-order system of the form (5.1) and (5.2) does not have a periodic orbit. The system defines a vector field $\mathbf{u} = (\dot{x}, \dot{y}) = (f(x, y), g(x, y))$. It is assumed that this vector field exists at all points in the plane, and is differentiable. Suppose that the divergence of this vector field has the same sign everywhere, either

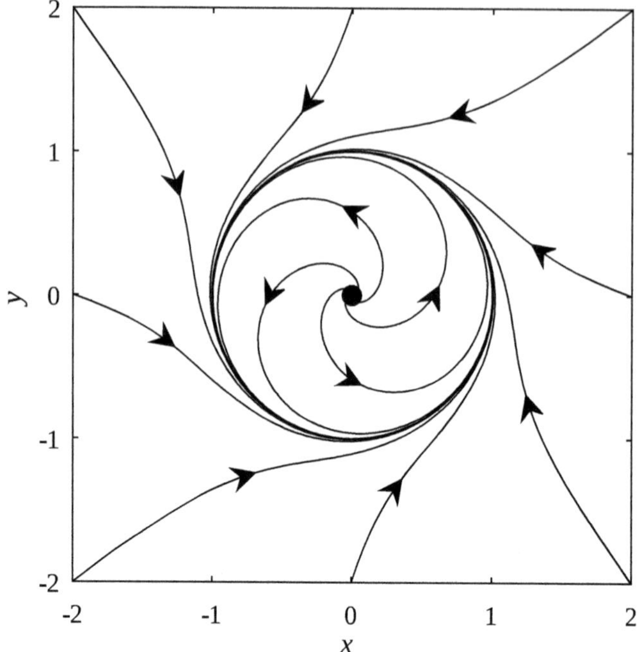

Fig. 5.8 Phase plane for (5.26), (5.27). The origin is an unstable spiral and there is a stable periodic orbit lying on the circle $x^2 + y^2 = 1$

positive or negative. Then the system (5.1) and (5.2) cannot have a periodic orbit. This result follows from the two-dimensional form of the divergence theorem of vector calculus. Recall that the divergence of the vector field $\mathbf{u}$ is a scalar field, defined by

$$\text{Div } \mathbf{u} = \nabla \cdot \mathbf{u} = \frac{\partial f}{\partial x} + \frac{\partial g}{\partial y}.$$

Suppose that the closed curve C is a periodic orbit. Then the vector field $\mathbf{u}$ points along the curve C at all points. However, the divergence theorem states that

$$\int_C \mathbf{u} \cdot \mathbf{n}\, ds = \iint_D \nabla \cdot \mathbf{u}\, dA, \tag{5.31}$$

where $\mathbf{n}$ is the unit outward normal vector, perpendicular to the curve C, ds is a line element along C, D is the region inside C, and dA is an area element within D. This leads to a contradiction, since the left-hand side of (5.31) is zero, as $\mathbf{u}$ and $\mathbf{n}$ are perpendicular if C is a periodic orbit, so the dot product is zero, but the right-hand side is either positive or negative. This result is often called the *Bendixson–Dulac Theorem* or *Bendixson's negative criterion*.

Intuitively, the divergence test says that if the vector field is expanding (or contracting) everywhere, it cannot form a closed periodic orbit.

Example 5.12 Show that there are no periodic orbits in the system

$$\dot{x} = x + y - y^2, \qquad \dot{y} = -x - x^3 - y^2 + y^3.$$

The divergence of the vector field in this example is $1 - 2y + 3y^2 = (1-y)^2 + 2y^2 > 0$, so by the divergence test, there is no periodic orbit.

An extended form of the divergence test states that if there is a scalar function $h(x, y)$, called a *weighting function*, such that $\nabla \cdot (h\mathbf{u})$ has the same sign everywhere, then there can be no periodic orbit.

5.8.2 The Curl Test and Gradient Systems

Suppose that for all x and y,

$$\frac{\partial g}{\partial x} - \frac{\partial f}{\partial y} = 0. \tag{5.32}$$

Then the system $(\dot{x}, \dot{y}) = \mathbf{u} = (f(x, y), g(x, y))$ has no periodic orbits.

In three dimensions, the above expression is the z component of the *curl* of the vector field $\mathbf{u}$. There are two ways to show this result. One is to apply Green's theorem. The second is to use the result that if a vector field $\mathbf{u}$ has zero curl at all points in the plane, it can be written as the gradient of some scalar field ϕ, that is,

$$\operatorname{curl} \mathbf{u} = 0 \;\Rightarrow\; \mathbf{u} = \nabla\phi = \left(\frac{\partial \phi}{\partial x}, \frac{\partial \phi}{\partial y}\right).$$

In this case, the system (5.1) and (5.2) is called a *gradient* system, or a *gradient-like* system. Now consider the rate of change of ϕ. Using the chain rule,

$$\dot{\phi} = \frac{\partial \phi}{\partial x}\dot{x} + \frac{\partial \phi}{\partial y}\dot{y} = \dot{x}^2 + \dot{y}^2 \geq 0.$$

So ϕ is always increasing, except at a fixed point. If there is a periodic orbit, then after one period of the orbit, the trajectory has returned to its original position, so ϕ must return to its original value. This is a contradiction, so no periodic orbit can exist.

This result also has a simple intuitive interpretation: a periodic orbit is a rotation, and the curl of a vector field is a measure of its rotation, so some non-zero curl is necessary for a periodic orbit to exist.

Example 5.13 Show that there are no periodic orbits in the system

$$\dot{x} = 2x - y + 2xy^2, \qquad \dot{y} = -x + 3y^2 + 2x^2y.$$

For this example, the quantity in (5.32) is $-1 + 4xy - (-1 + 4xy) = 0$, so the curl of the vector field is zero and there cannot be any periodic orbits. The function $\phi(x, y)$ in this case is $\phi = x^2 - xy + y^3 + x^2y^2$.

As for the divergence test, there is a generalised form of the curl test. If there exists a scalar function $h(x, y)$ that has the same sign everywhere in the plane, such that

$$\frac{\partial(hg)}{\partial x} - \frac{\partial(hf)}{\partial y} = 0, \tag{5.33}$$

then there can be no periodic orbit.

5.8.3 The Index Test

Let C be any closed curve in the phase plane (not, in general, periodic orbit), that does not pass through a fixed point. Then the *index* of the curve is defined as follows. Start from any point on the curve, and measure the angle that the flow field vector $\mathbf{u} = (f(x, y), g(x, y))$ makes with, say, the unit vector in the x direction. This angle is clearly defined, since the curve does not pass through any fixed points. Then follow the curve around in an anticlockwise direction, monitoring this angle, and return to the original position. The angle at that point must be the same, but in the process, the angle has changed by an integer multiple (positive, zero or negative), of 2π. This multiple is known as the *index* of the curve.

Now suppose that the curve is perturbed slightly. Since the index is an integer function of the curve, it should, in general, not change. The index of a curve can only change if the curve is moved in such a way that it crosses a fixed point. So the index of a closed curve depends on the fixed points inside it. If the curve contains no fixed points, its index is zero. If it contains a node or a spiral (stable or unstable), but no other fixed points, the index is $+1$. If it contains a saddle point only, its index is -1. Therefore we can also define the index of a fixed point, as the index of a closed curve that encloses that fixed point but no other fixed points. Spirals, nodes and centres have index $+1$, while saddles have index -1. The index of a curve containing more than one fixed point is simply the sum of the indices of those points. Figure 5.9 shows a phase plane with a saddle, a node and a spiral, with circular curves around each point to illustrate the index of each point.

Now suppose that the curve C is a periodic orbit. Then the vector field points along the curve at all points, so its index is $+1$. This is only possible if the sum of the indices of the fixed points inside C is $+1$.

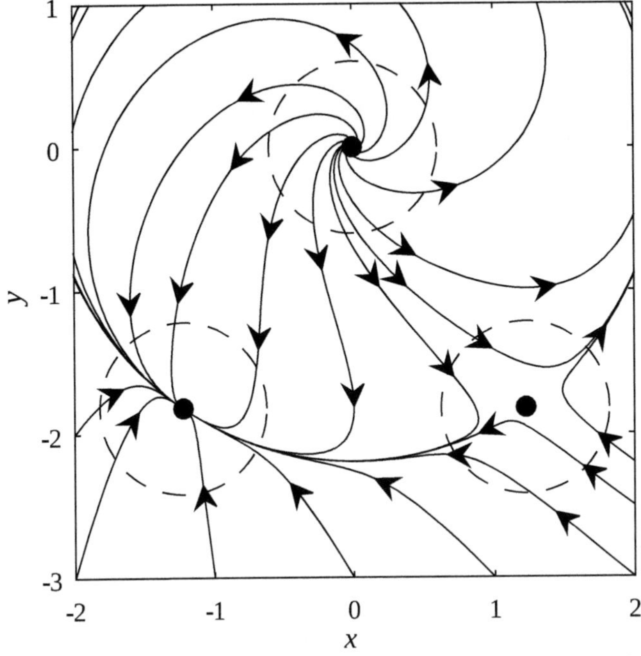

Fig. 5.9 This phase plane has an unstable spiral, a stable node and a saddle. Following the dashed circles around either the node or the spiral, the direction of the trajectories rotates once in the same direction as the circle is traversed, so the index is +1. For the saddle, the trajectories rotate once in the opposite direction, so the index is −1

Considering the index of the fixed points of a system gives a number of results about periodic orbits, which can be used to rule out periodic orbits in some cases:

- Every periodic orbit must contain at least one fixed point.
- If a periodic orbit contains one fixed point, that fixed point must be a node, spiral or centre, not a saddle.
- If a periodic orbit contains several fixed points, the sum of the indices of these points must be + 1. So, for example, a periodic orbit can enclose two nodes and a saddle, but not two saddles and a node.

Example 5.14   Show that there are no periodic orbits in system

$$\dot{x} = x + xy^2, \qquad \dot{y} = -y + x^3 - xy^3.$$

Since $\dot{x} = x(1+y^2)$, any fixed point must have $x = 0$, and then from the $\dot{y}$ equation, y must also be zero. So $(0, 0)$ is the only fixed point. It is a saddle, with index − 1, so there can be no periodic orbits.

5.8.4 *The Invariant Line Test*

The existence of invariant lines or curves alone does not disprove the existence of periodic orbits, but it does constrain the behaviour of the system significantly, and when combined with some of the previous methods can sometimes be used to rule out periodic orbits. The key point is that no trajectory can cross an invariant line, and this limits the possible locations of periodic orbits. This is illustrated with three simple examples.

Example 5.15 Use the divergence test combined with the existence of invariant lines to show that there are no periodic orbits in the system

$$\dot{x} = x - x^2 y, \qquad \dot{y} = -y + 2y^2 x.$$

The divergence is $1 - 2xy - 1 + 4yx = 2xy$. This does not always have the same sign, so the divergence test alone does not help. However, since $x = 0 \Rightarrow \dot{x} = 0$ and $y = 0 \Rightarrow \dot{y} = 0$, the lines $x = 0$ and $y = 0$ are invariant. So no trajectories can cross the axes, and these two invariant lines divide the phase plane into four disconnected regions. Now the divergence test can be applied in each of the four quadrants of the plane. For $x > 0$, $y > 0$, the divergence is always positive, so no periodic orbit can exist in this region, and no trajectory can leave this region. Similarly, the divergence has a fixed sign in each of the other three quadrants.

Example 5.16 By finding the invariant lines that pass through the origin, show that there are no periodic orbits in the system

$$\dot{x} = x - x^2 + 2y^2, \qquad \dot{y} = y + xy.$$

There are four fixed points, at $(0, 0)$, $(1, 0)$, $(-1, \pm 1)$. Consider a straight line through the origin, $y = mx$. The line is invariant if $\dot{y} = m\dot{x}$ on $y = mx$, so $mx + mx^2 = mx - mx^2 + 2m^3 x^2$, which simplifies to $m = m^3$. Hence there are three invariant lines, $y = 0$, $y = x$ and $y = -x$. Each fixed point lies on at least one of the invariant lines. Using the index test, any periodic orbit must enclose one of the fixed points, but then it would have to cross an invariant line, which is not possible, so there are no periodic orbits.

Example 5.17 Show that the SIR model of Example 5.10 has no periodic orbits.

In the SIR model, there are infinitely many fixed points along the line $I = 0$ but no other fixed points. But this line is an invariant line, so no trajectory can cross it. Since the index test shows that every periodic orbit must enclose a fixed point, a periodic orbit cannot exist because it would have to cross the invariant line.

5.8.5   *Existence of Periodic Orbits*

Having listed several methods that can be used to show that periodic orbits do not exist, this section ends with a theorem that can be used in some cases to show that they do exist.

Theorem 5.2 (Poincaré–Bendixson) *Let D be a closed, bounded subset of the plane that contains no fixed points. Suppose that any trajectory that starts in D or on the boundary of D remains within D for all time, so that no trajectory can leave D. Then D contains at least one stable periodic orbit.*

Recall from Sect. 5.8.3 that every periodic orbit must contain at least one fixed point. Since the theorem states that D does not contain a fixed point, the domain D cannot be simply connected. In most cases, D is an annulus-shaped region bounded by an inner closed curve C_1 and an outer closed curve C_2 (see Fig. 5.10).

The theorem works because trajectories cannot leave D, and trajectories cannot cross. Any trajectory in D remains in D, and must keep moving as there are no fixed points in D. This is only possible if the trajectory approaches a periodic orbit.

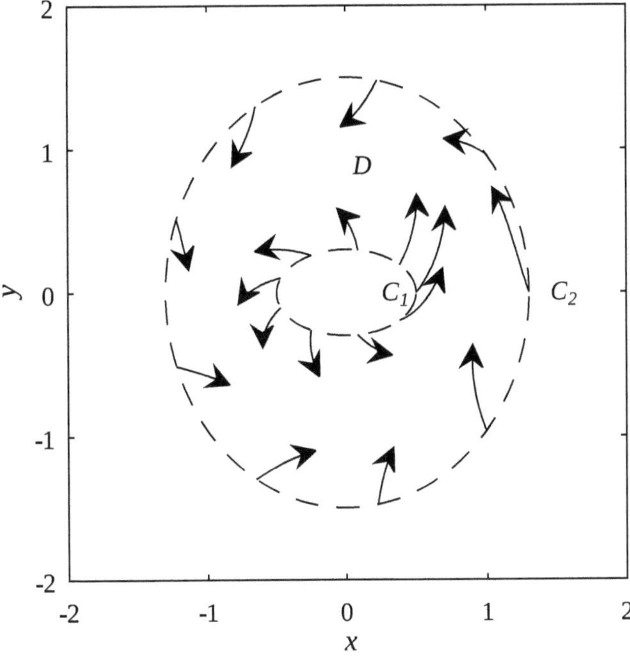

Fig. 5.10 The Poincaré–Bendixson theorem. The region D is bounded by the inner and outer dashed curves C_1 and C_2. All trajectories on C_1 and C_2 point into D and there are no fixed points in D

Note that there may be more than one periodic orbit in D. For example, there could be two stable periodic orbits and one unstable one.

There is an alternative version of the theorem, in which all trajectories point outward on the boundary of D, in which case the result is that there exists an unstable periodic orbit in D.

A simple but pointless application of the Poincaré–Bendixson theorem is provided by Example 5.11, Fig. 5.8. In that example, $\dot{r} > 0$ for $r < 1$ and $\dot{r} < 0$ for $r > 1$. The curve C_1 can be chosen to be the circle $r = r_-$, for any $r_- < 1$, and C_2 can be set as be the circle $r = r_+$, for some $r_+ > 1$. The conditions of the theorem are then satisfied so there exists a periodic orbit between the two circles. However, there is no need to use the theorem in this case as the periodic orbit can be found explicitly.

In many examples, the theorem can be applied by using polar coordinates, and then using inequalities to find suitable choices for C_1 and C_2, as in the following example.

Example 5.18 Show that there is a periodic orbit in the system

$$\dot{x} = x - y - 2x^3 - 2y^3, \tag{5.34}$$

$$\dot{y} = x + y + 2x^3 - 2y^3. \tag{5.35}$$

First, check for fixed points. Adding the two equations, $2x - 4y^3 = 0$ at a fixed point, and subtracting them, $2y + 4x^3 = 0$. Eliminating, say, y, leads to $x = -16x^9$, so either $x = 0$ or $16x^8 = -1$, which has no real solutions. So the origin is the only fixed point. The matrix of the linearised system at the origin has trace 2 and determinant 2, so the origin is an unstable spiral.

Now use polar coordinates, as in Example 5.11. Using (5.28),

$$
\begin{aligned}
r\dot{r} &= x\dot{x} + y\dot{y} \\
&= r^2 - 2(x^4 + y^4) - 2xy^3 + 2yx^3 \\
&= r^2 - 2(r^4 - 2x^2y^2) + 2xy(x^2 - y^2) \\
&= r^2 - 2r^4(1 - 2\cos^2\theta \sin^2\theta) + 2r^4 \cos\theta \sin\theta(\cos^2\theta - \sin^2\theta) \\
&= r^2 - 2r^4\left(1 - \frac{1}{2}\sin^2 2\theta\right) + r^4 \sin 2\theta \cos 2\theta \\
&= r^2 - 2r^4\left(1 - \frac{1}{4}(1 - \cos 4\theta)\right) + \frac{1}{2}r^4 \sin 4\theta \\
&= r^2 - \frac{r^4}{2}(3 + \cos 4\theta - \sin 4\theta).
\end{aligned}
$$

The term in the bracket is bounded above by 5 and bounded below by 1. Then

$$\dot{r} > r - 5r^3/2 = r(1 - 5r^2/2) \quad \text{and} \quad \dot{r} < r - r^3/2 = r(1 - r^2/2).$$

So if $r^2 < 2/5$, $\dot{r} > 0$ and if $r^2 > 2$, $\dot{r} < 0$. These two circles define the annular region D needed for the Poincaré–Bendixson theorem. On the inner circle $r = \sqrt{2/5}$ trajectories point outwards, into D, and on the outer circle $r = \sqrt{2}$ they point inwards, again into D. Therefore no trajectories can escape from the region D, and there must be a periodic orbit lying within this region.

The final example of this chapter brings together several of the topics discussed previously, and provides an example of a fixed point with unusual stability properties.

Example 5.19 Find and classify the fixed points of the system

$$\dot{x} = x(1 - x^2 - y^2) - y + xy, \tag{5.36}$$

$$\dot{y} = y(1 - x^2 - y^2) + x - x^2. \tag{5.37}$$

Find an invariant curve, and state whether there is a periodic orbit. Sketch the phase plane and describe the behaviour of trajectories as $t \to \infty$. Are there any fixed points that are Lyapunov stable or asymptotically stable?

Given the similarity to Example 5.11, consider the system in polar coordinates. Following the method of that example,

$$r\dot{r} = x\dot{x} + y\dot{y} = r^2(1 - r^2) + x(-y + xy) + y(x - x^2) = r^2 - r^4,$$

Also

$$r^2\dot{\theta} = x\dot{y} - y\dot{x} = x^2 - x^3 + y^2 - xy^2 = r^2 - xr^2 = r^2 - r^3\cos\theta,$$

so in polar coordinates,

$$\dot{r} = r - r^3, \qquad \dot{\theta} = 1 - r\cos\theta.$$

In polar coordinates the fixed points are $r = 0$ and $r = 1$, $\theta = 0$, which in Cartesian coordinates are $x = y = 0$ and $x = 1$, $y = 0$. The origin is an unstable spiral, with an anticlockwise rotation since $\dot{\theta} > 0$ for small r. The stability of the fixed point $(1, 0)$ can be investigated in the usual way, by finding the Jacobian, which is

$$J_{(1,0)} = \begin{pmatrix} -2 & 0 \\ -1 & 0 \end{pmatrix}.$$

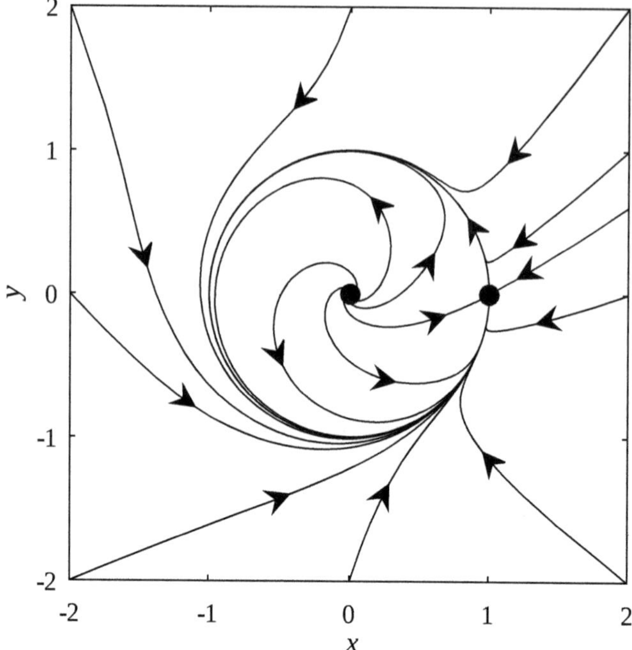

Fig. 5.11 Phase plane for Example 5.19. Trajectories approach the fixed point at $(1, 0)$. However, this point is neither asymptotically stable nor Lyapunov stable

One eigenvalue is -2 with eigenvector $(2, 1)^T$. The other is 0, with eigenvector $(0, 1)^T$. Recall from Example 5.2 that this means that in the linearised system, there is a line of fixed points, extending in the y direction from $(1, 0)$. But this is not the case in the nonlinear system.

Since the r equation is $\dot{r} = r - r^3$, the curve $r = 1$ is an invariant curve, because $\dot{r} = 0$ when $r = 1$. However, unlike Example 5.11, $r = 1$ is not a periodic orbit, because there is a fixed point at $r = 1, \theta = 0$. On the circle $r = 1, \dot{\theta} = 1 - \cos\theta \geq 0$, with equality only at $\theta = 0$.

We can now describe the behaviour and sketch the phase plane. For any initial condition apart from $r = 0, r \to 1$ as $t \to \infty$. Trajectories on the unit circle rotate anticlockwise and approach the fixed point at $(1, 0)$. The phase plane is shown in Fig. 5.11.

Now consider the stability of the fixed points. The point at the origin is an unstable spiral. The point at $(1, 0)$ has the property that all trajectories, except for the initial condition $(0, 0)$, approach it as $t \to \infty$. But this does not mean that the fixed point is asymptotically stable, since asymptotic stability requires Lyapunov stability. This point is not Lyapunov stable, because trajectories that start near the fixed point do not necessarily stay near it. A trajectory that starts with a small perturbation in the positive y direction takes a long detour around the unit circle before returning to approach $(1, 0)$. So both the fixed points are unstable.

5.9 Systems of Higher Order

Many of the methods of this chapter can be extended to higher order systems of differential equations. In an n-th order system, fixed points are defined in the same way, and their stability depends on the eigenvalues of the $n \times n$ Jacobian matrix. If all the eigenvalues have negative real part, the fixed point is stable. The phase plane becomes an n-dimensional phase space.

Fixed points can be described in a similar way to the classification used in this chapter. A fixed point is a stable node if all of its eigenvalues are real and negative, and an unstable node if they are all real and positive. If the eigenvalues are real but have different signs, the fixed point is described as a saddle.

Sketching the phase space is difficult in n dimensions, but it can be helpful to plot two-dimensional cross-sections. This is particularly useful if the system has a symmetry. For example, suppose that a third-order system for three dependent variables x, y and z has the reflection symmetry $x \rightarrow -x$. Then, following the argument of Sect. 5.6, no trajectory can cross the plane $x = 0$. So this plane divides the three-dimensional phase space into two symmetry-related halves, and if the initial condition lies in the plane $x = 0$, the solution trajectory must remain in this plane for all t. In this case we can sketch the two-dimensional y, z phase plane. This generalisation of an invariant line is called an invariant plane, and in n dimensions it is described as an invariant subspace of dimension $n - 1$.

Hamiltonian systems can occur if n is even. The generalised form of the Hamiltonian equations (5.20) is

$$\dot{x}_i = \frac{\partial H}{\partial y_i}, \qquad \dot{y}_i = -\frac{\partial H}{\partial x_i},$$

for $1 \leq i \leq n/2$, from which it follows that H is a conserved quantity.

Periodic orbits commonly occur in higher order systems. However, the Poincaré–Bendixson theorem does not extend to systems of order higher than two, because in higher dimensions, trajectories can wind around each other without crossing.

There are various new types of behaviour that can occur for $n > 2$. The most important new possibility that occurs for $n > 2$ is the phenomenon of chaos, which will be discussed in Chap. 8.

Key Points from Chap. 5

- Most second-order nonlinear differential equations cannot be solved analytically.
- Applications include models for competition between species, problems arising from Newton's second law of motion, and modelling epidemics.
- Any second-order differential equation can be written as two coupled first-order ones, $\dot{x} = f(x, y)$, $\dot{y} = g(x, y)$.
- The qualitative approach, based on plotting trajectories in the (x, y) *phase plane*, can be used to describe the solutions of the system, in particular the behaviour as $t \to \infty$.
- A *fixed point* is a point in the plane where f and g are both zero.
- The behaviour of the system near a fixed point is determined by the *eigenvalues* of the *Jacobian matrix* evaluated at the fixed point.
- The stability of each fixed point and its type can be found using the methods of Chap. 4. However, this analysis only applies near the fixed point.
- Having analysed the type of each of fixed point, the complete phase plane picture can be sketched using various techniques, including the following:
- A *nullcline* is a curve in the plane on which either f or g is zero.
- Trajectories can never cross, although they may meet at a fixed point.
- If a system has *Hamiltonian form*, with a Hamiltonian function $H(x, y)$, then trajectories are the contour lines $H = $ constant and the only possible types of fixed points are saddles and centres.
- An *invariant line* or *invariant curve* has the property that if the initial condition lies on the curve, the trajectory remains on that curve for all t.
- Second-order systems can have *periodic orbits*, closed curves in the phase plane representing periodic solutions. Usually they cannot be found analytically. In some cases it can be proved that they exist or that they do not exist.

Exercises

5.1 Find the nullclines of the system

$$\dot{x} = (x - 1)(y^2 - 4),$$
$$\dot{y} = x^2 - y^2,$$

and sketch them. Use the nullclines to find the location of all the fixed points. What happens if the first equation is changed to $\dot{x} = (x - 1)(y^2 - 1)$?

5.2 Investigate the type of the fixed points in Example 5.3. Are your results consistent with the type suggested by the arrows on the nullclines shown in Fig. 5.2?

5.3 (a) Show that the system

$$\dot{x} = y, \qquad \dot{y} = x(x^2 - 1)$$

is Hamiltonian, and find the Hamiltonian function $H(x, y)$. Check that H is constant on any trajectory.
(b) Find the three fixed points. What is the value of H at each fixed point?
(c) Find the equations of two curves on which H is constant that pass through two of the fixed points. Sketch these curves in the (x, y) plane and find the direction of trajectories on them.
(d) Hence sketch the phase plane (you do not need to investigate the Jacobian at each fixed point in order to draw the phase plane).

5.4 Consider the system of Example 5.6.

(a) Show that as well as the invariant curve $y = x^2$, there is an invariant line.
(b) Find all the fixed points of the system. How many of them lie on the invariant line, and how many on the invariant curve? Find the direction of trajectories that lie on the invariant line or the invariant curve.
(c) Show that all of the fixed points are unstable, and sketch the phase plane.

5.5 Find the three invariant lines of the system

$$\dot{x} = x + xy - 4x^3,$$
$$\dot{y} = y + 4x^2 - y^3.$$

Hint: find one invariant line by factorising one of the equations. For the other two, try a straight line through the origin.

5.6 Following the method of Example 5.7, show that for any second-order system with the reflection symmetry $x \leftrightarrow -x$, $x = 0$ is an invariant line. What conclusion can be drawn if the system has both $x \leftrightarrow -x$ and $y \leftrightarrow -y$ symmetry?

5.7 Consider the Lotka–Volterra model as in Example 5.8 but with a change to the growth rate of the x species,

$$\dot{x} = 5x - x^2 - xy,$$
$$\dot{y} = 4y - 2y^2 - xy.$$

Find and classify the fixed points and sketch the phase plane. Discuss how the behaviour differs from the case of Example 5.8.

5.8 For the general Lotka–Volterra model (5.3), (5.4), investigate whether there are parameter values for which both of the fixed points corresponding to only one species existing are stable.

5.9 The following model describes the interaction of two species, one of which is a predator and the other is its prey.

$$\dot{x} = 3x - x^2 - xy,$$
$$\dot{y} = -y + xy.$$

(a) Which variable is the predator and which is the prey? Explain the signs of the nonlinear terms in the system.
(b) Find the fixed points and investigate their stability.
(c) Sketch the phase plane and describe the qualitative behaviour of the system in terms of the abundance of each species as a function of time.

5.10 A modification of the predator–prey model of question (5.9) is

$$\dot{x} = x - xy,$$
$$\dot{y} = -y + xy,$$

for $x, y > 0$.

(a) Show that the system is not Hamiltonian.
(b) Define $u = \log x$, $v = \log y$ and obtain differential equations for u and v.
(c) Show that the (u, v) system is Hamiltonian, and find the Hamiltonian function $H(u, v)$. Find the only fixed point in the u, v plane and determine its type.
(d) Deduce a relationship between x and y on any trajectory in the original system.
(e) Obtain the same relationship by dividing $\dot{y}$ by $\dot{x}$ and solving the differential equation for dy/dx.
(f) Without doing any further work, sketch the x, y phase plane and describe how the behaviour of the system differs from the system in Exercise 5.9.

5.11 Consider the damped pendulum of Example 5.9 for arbitrary $\beta > 0$. Do the types of the fixed points at $(0, 0)$ and $(\pi, 0)$ remain the same for all values of β? If not, find the value of β where there is a change.

5.12 In the SIR model of Sect. 5.1.3 and Example 5.10, people who have recovered are assumed to be immune, so they cannot be reinfected.

(a) Modify the model (5.25) so that those who have recovered, R, can become susceptible again at a rate cR. Check that the total population is constant in your model.
(b) Eliminate R to obtain a second-order system.
(c) Investigate the existence of fixed points in this system.
(d) For the parameter values $a = 4$, $b = 2$, $c = 1$, find the type of each fixed point, and hence sketch the phase plane.
(e) Interpret the results, comparing this model with the case studied in Example 5.10.

5.13 Derive the equations (5.28) by writing r and θ in terms of x and y and then differentiating.

5.14 Consider the problem of Example 5.11 but with some changes of sign,

$$\dot{x} = -x - y + x(x^2 + y^2),$$
$$\dot{y} = -y + x + y(x^2 + y^2).$$

Express this system in polar coordinates and hence describe the behaviour.

5.15 Use the divergence test with the weighting function $h(x, y) = 1/xy$ to show that there are no periodic orbits in the Lotka–Volterra system of Example 5.8 (assume that $x > 0$, $y > 0$).

5.16 Show that there are no periodic orbits in the system of Example 5.6 investigated in Exercise 5.4.

5.17 Show that the equation

$$\ddot{u} + (4u^2 + \dot{u}^2 - 1)\dot{u} + u = 0$$

has a periodic orbit.

5.18 Investigate whether there are any periodic orbits in the following systems. There is one example of each of the methods in Sects. 5.8.1–5.8.5.

(a) $\dot{x} = \cos x \sin y$, $\quad \dot{y} = \sin x \cos y$.
(b) $\dot{x} = x^2 - 1$, $\quad \dot{y} = x(y^2 - 1)$.
(c) $\dot{x} = x - x(x^2 + y^2) - y^3$, $\quad \dot{y} = y - y(x^2 + y^2) + x^3$.
(d) $\dot{x} = x/(1 + y^2) + y^3 - 2y^4$, $\quad \dot{y} = x + 2y + x^2 y + \sin x + y^5$.
(e) $\dot{x} = 1 - x^2$, $\quad \dot{y} = 1 + xy$.

Chapter 6
Bifurcations

Most mathematical models include parameters. For example, in the Lotka–Volterra model introduced in Sect. 5.1.1, there are several parameters, representing the growth rates of the two species and the competition between them. Figure 5.5 shows the phase plane diagram for a particular choice of these parameters. Now suppose that a small change is made to one of these parameters. Since the phase plane diagram is only a qualitative picture of the behaviour of the system, this small change usually makes no significant change to the picture. However, Exercises 5.7 and 5.8 show that the behaviour of the Lotka–Volterra system does depend on the parameter values. As a parameter varies, the qualitative behaviour remains the same, except at certain points known as *bifurcation points*. The system is said to have a *bifurcation* as the parameter passes through one of these special values where the picture changes.

This definition of a bifurcation, as a point in parameter space at which there is a qualitative change in the behaviour of the system, is too vague to be able to work with mathematically. But it is the only definition that covers all the things that are known as bifurcations. These include a fixed point becoming unstable, or the number of fixed points changing, or a periodic orbit appearing, disappearing or becoming unstable. More precise, mathematical definitions can be given for specific types of bifurcation.

Bifurcation theory provides a qualitative description of the behaviour of a system of nonlinear differential equations not just for all initial conditions, as in Chap. 5, but also for all possible values of the parameters.

Remarkably, it turns out that there are only a few types of bifurcation that commonly occur. Each one has its own *normal form*, a simplified form that it can be reduced to near the bifurcation point. Furthermore, the same types of bifurcations that occur in simple first-order differential equations as a parameter is varied also occur in systems of $n > 1$ differential equations. This surprising fact occurs because very near a bifurcation point, the dimension can often be reduced from n to 1 by a

P. C. Matthews, *Differential Equations, Bifurcations and Chaos*,
Springer Undergraduate Mathematics Series,
https://doi.org/10.1007/978-3-031-99543-9_6

process described in Sect. 6.2.1. In view of this fact, it makes sense to start by looking at the first-order case.

Remark 6.1 This chapter goes into some fairly detailed aspects of bifurcation theory. For a lighter version of the topic, omit the detailed derivation via Taylor series of the three main types of bifurcation in Sect. 6.1, and focus on their normal forms and bifurcation diagrams. Similarly, in Sect. 6.2, omit the computations needed to reduce the order, but look at the phase plane diagrams such as Figs. 6.5 and 6.6 to understand the concept. In Sect. 6.2.3, study the figures illustrating global bifurcations and their descriptions, but omit the detailed algebraic calculations. In short, if you find this chapter too hard, just look at the pictures.

6.1   Bifurcations in First-Order Differential Equations

Consider a first-order, autonomous differential equation for $x(t)$ that includes a parameter μ,

$$\frac{dx}{dt} = f(x, \mu), \tag{6.1}$$

and assume that f is infinitely differentiable. Bifurcation theory is concerned with how the behaviour of this equation changes as the parameter μ is varied. For a fixed value of μ, solutions can be described qualitatively as in Chap. 3, by finding the fixed points and investigating their stability, then drawing a phase line to show the behaviour of solutions.

Now suppose that a fixed point x_0 of (6.1) has been found for a certain value of μ, $\mu = \mu_0$, so

$$f(x_0, \mu_0) = 0.$$

What happens to this fixed point as the parameter μ is varied? Can we write x as a function of μ? If x and μ are close to the known values (x_0, μ_0), so $x = x_0 + \Delta x$, $\mu = \mu_0 + \Delta \mu$, where Δx and $\Delta \mu$ are small, then the rule for small changes of a function of two variables, see (A.7) in the Appendix, is

$$f(x, \mu) \approx f(x_0, \mu_0) + \Delta x \frac{\partial f}{\partial x} + \Delta \mu \frac{\partial f}{\partial \mu},$$

where the partial derivatives are evaluated at (x_0, μ_0) and the error terms that make this an approximation rather than an equality are quadratic in the small quantities Δx, $\Delta \mu$ or smaller. At a fixed point, $f(x, \mu) = 0$, so the small change in the location of the fixed point corresponding to the small change in the value of μ is

$$\Delta x \approx -\Delta \mu \frac{\partial f}{\partial \mu} \bigg/ \frac{\partial f}{\partial x}.$$

Therefore, the fixed point can be followed as μ varies, to define a local function $x(\mu)$, provided that $\frac{\partial f}{\partial x} \neq 0$. This process is known as *continuation*, or more informally as following a *branch* of fixed points. The argument above forms the basis of the following theorem.

Theorem 6.1 (Implicit Function Theorem) *Suppose that $f(x, \mu)$ is continuously differentiable, with $f(x_0, \mu_0) = 0$. Then if $\frac{\partial f}{\partial x}(x_0, \mu_0) \neq 0$, the equation $f(x, \mu) = 0$ defines a unique continuous function $x(\mu)$ for μ sufficiently close to μ_0.*

The theorem is called the Implicit Function Theorem (IFT) because it gives the condition under which the equation $f(x, \mu) = 0$ implicitly defines a function $x(\mu)$. A simple example helps to clarify this.

Example 6.1 To illustrate the IFT, consider the differential equation

$$\dot{x} = 4x^2 + \mu^2 - 100. \tag{6.2}$$

Fixed points lie on the ellipse $4x^2 + \mu^2 = 100$, see Fig. 6.1. This equation alone does not define a function $x(\mu)$, because a function must have a uniquely defined value, but when $\mu = 0$ for example, we don't know whether to choose the value $x = 5$ or $x = -5$. But suppose that we start from the fixed point $x = 5$ at $\mu = 0$, labelled A in the figure. Then making small changes to μ traces out a curve on the upper half of the ellipse, making a uniquely defined continuous function $x(\mu)$, near to the starting point. Similarly, starting from the fixed point B, at $x = -4$, $\mu = 6$, if μ varies, there is a well-defined nearby curve $x(\mu)$.

This works for any starting point, except for the end points C and D of the ellipse, where $x = 0$, $\mu = \pm 10$. Starting at the fixed point C, where $x = 0$ and $\mu = 10$, we cannot continue the fixed point to larger values of μ because no solutions exist

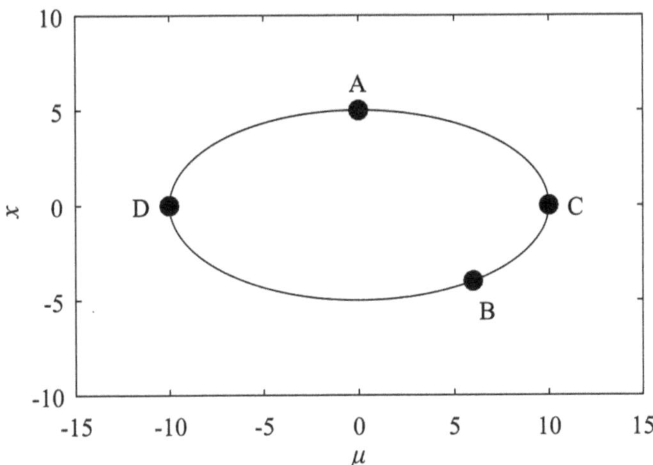

Fig. 6.1 The Implicit Function Theorem for (6.2). The elliptical curved line shows the fixed points. Starting from the fixed points A or B, a unique smooth curve of fixed points can be followed as μ varies. But this is not true at points C and D

for $\mu > 10$. Nor can we decrease μ, because we don't know whether to follow the upper or lower branch, so there is no uniquely defined function $x(\mu)$.

This is what the implicit function theorem says: the function $x(\mu)$ can be defined near $x = x_0$, $\mu = \mu_0$ if $\frac{\partial f}{\partial x} = 8x \neq 0$, so we can track the fixed point from a starting point x_0 if $x_0 \neq 0$, which is equivalent to $\mu \neq \pm 10$.

Note that the IFT does not work both ways. If $\frac{\partial f}{\partial x}(x_0, \mu_0) \neq 0$, the theorem says that there is a unique, nearby solution $x(\mu)$. If $\frac{\partial f}{\partial x}(x_0, \mu_0) = 0$, then usually, as in Example 6.1, it is not possible to vary μ and obtain a function $x(\mu)$ locally, although in some cases this may be possible. An example is the equation $f(x, \mu) = \mu - x^3 = 0$. Here, $\frac{\partial f}{\partial x}(0, 0) = 0$, so the IFT tells us nothing, but in fact there is a unique function $x(\mu) = \mu^{1/3}$ defined for all values of μ.

Clearly, the change that occurs in Example 6.1 at $\mu = \pm 10$, where the number of fixed points changes from two to zero, satisfies the general definition of a bifurcation given at the start of this chapter: a value of μ at which there is a qualitative change in the behaviour. This inspires the following definition of a bifurcation for a first-order differential equation.

Definition 6.1 The first-order differential equation $\dot{x} = f(x, \mu)$ with a real parameter μ has a *bifurcation* at the point $x = x_0$, $\mu = \mu_0$ if $f(x_0, \mu_0) = 0$ and $\frac{\partial f}{\partial x}(x_0, \mu_0) = 0$, provided that $\frac{\partial f}{\partial x}$ changes sign as the branch of fixed points is followed through this point.

A bifurcation occurs when the condition needed for the IFT to guarantee the existence of a local implicitly defined function does not hold. The extra condition that the partial derivative should change sign is needed to rule out examples such as that mentioned above, $\dot{x} = \mu - x^3$, where the partial derivative is zero at the origin but there is no qualitative change. However, this extra condition is satisfied except in a few special cases.

Another interpretation of this definition for a bifurcation in a first-order differential equation, from Sect. 3.2.1, is that a bifurcation occurs when a fixed point crosses the borderline between stability and instability.

Definition 6.1 can be used to derive systematically the three types of bifurcation that commonly occur in first-order differential equations. This is the aim of the following three sections. Each bifurcation has its own *normal form*, a simple form that it can always be turned into by rescaling the variables, sufficiently close to the bifurcation. The first type of bifurcation is that seen in Fig. 6.1, where two fixed points merge and disappear.

6.1.1 Saddle–Node Bifurcation

Suppose that the differential equation (6.1) has a bifurcation at $x = x_0$, $\mu = \mu_0$. What is the behaviour of the equation near this point? This question can be answered by using a Taylor series expansion of the function $f(x, \mu)$ near the bifurcation point (recall that it is assumed that f is infinitely differentiable, so the Taylor series exists). To avoid some cumbersome expressions, suppose that $x_0 = 0$ and $\mu_0 = 0$, so the bifurcation occurs at the origin of the x, μ plane. This can always be achieved by re-defining the coordinates to shift the bifurcation point. The Taylor series of (6.1) near $x = x_0 = 0$, $\mu = \mu_0 = 0$ up to quadratic terms is

$$\dot{x} = f(x, \mu) = f(0,0) + x f_x + \mu f_\mu + \frac{1}{2} x^2 f_{xx} + x \mu f_{x\mu} + \frac{1}{2} \mu^2 f_{\mu\mu} + O(3). \quad (6.3)$$

See Sect. A.1 in the Appendix for explanation of the $O(3)$ order notation that represents cubic and smaller terms, and Sect. A.5 for Taylor series for functions of two variables. Here, the shorthand notation f_x, f_μ is used for the partial derivatives, which are all evaluated at $x = 0$, $\mu = 0$, so f_{xx} and all the other derivatives are just known numbers. Since we are looking near the bifurcation point, x and μ are small, so the neglected cubic terms should be very small and the truncated Taylor series should give a good approximation. If there is a bifurcation at $(0, 0)$, then according to Definition 6.1, $f(0, 0) = 0$ and $f_x(0, 0) = 0$, so the Taylor expansion simplifies to

$$\dot{x} = \mu f_\mu + \frac{1}{2} x^2 f_{xx} + x \mu f_{x\mu} + \frac{1}{2} \mu^2 f_{\mu\mu} + O(3). \quad (6.4)$$

Now consider the size of the terms in this equation, assuming that x and μ are small. Since $x\mu \ll \mu$ and $\mu^2 \ll \mu$, the third and fourth terms are small compared with the first one, as long as $f_\mu \neq 0$. But the second term cannot be ignored, because although x^2 is small, we do not know the relative magnitude of μ and x. The first two terms are the same order of magnitude if $\mu = O(x^2)$, so $x = O(\sqrt{\mu})$, as x and μ tend to 0, and with this scaling, the dominant terms are

$$\dot{x} = \mu f_\mu + \frac{1}{2} x^2 f_{xx}.$$

This equation can be simplified by rescaling $x = a\hat{x}$ and $\mu = c\hat{\mu}$, leading to

$$a\dot{\hat{x}} = c\hat{\mu} f_\mu + \frac{1}{2} a^2 \hat{x}^2 f_{xx}.$$

Now choose $a = -2/f_{xx}$ and $c = a/f_\mu$. Then, dropping the hat notation,

$$\dot{x} = \mu - x^2. \quad (6.5)$$

This equation is known as the *normal form* of the bifurcation. This means that it can always be written in this form after a rescaling of the variables. In (6.5) there are no fixed points for $\mu < 0$. For $\mu > 0$ there are two fixed points, $x = \pm\sqrt{\mu}$. Their stability depends on the sign of the x-derivative, $-2x$, so the solution $x = \sqrt{\mu}$ is stable and $x = -\sqrt{\mu}$ is unstable.

The bifurcation in (6.5) is known as a *saddle–node* or *turning point* bifurcation. It is called a saddle–node bifurcation because when it occurs in a second-order system, the two fixed points that merge and disappear at $\mu = 0$ are a saddle and a node, as will be seen in Sect. 6.2.

Any bifurcation can be illustrated using a *bifurcation diagram*: a plot of x as a function of μ showing all the fixed points that exist. A convention commonly used in bifurcation diagrams is that stable fixed points are shown as solid lines and unstable ones as dashed lines. The bifurcation diagram for (6.5) is shown in Fig. 6.2. For any value of μ, a vertical cross-section through the bifurcation diagram gives the phase line of the equation for that value of μ. Arrows can be drawn vertically upwards or downwards on the bifurcation diagram, depending on the sign of $\dot{x}$, as shown in Fig. 6.2, but these arrows are not necessary, as their direction can be deduced from the solid and dashed lines.

The choice of signs in the normal form (6.5) is just a convention. We could have chosen to define the normal form as

$$\dot{x} = \mu + x^2 \quad \text{or} \quad \dot{x} = -\mu + x^2 \quad \text{or} \quad \dot{x} = -\mu - x^2.$$

These can all be transformed into (6.5) by changing the sign of x and/or μ.

Note that in the rescaling used to convert the bifurcation into its normal form, we had to divide by f_{xx} and f_μ. So necessary (and in fact sufficient) conditions for

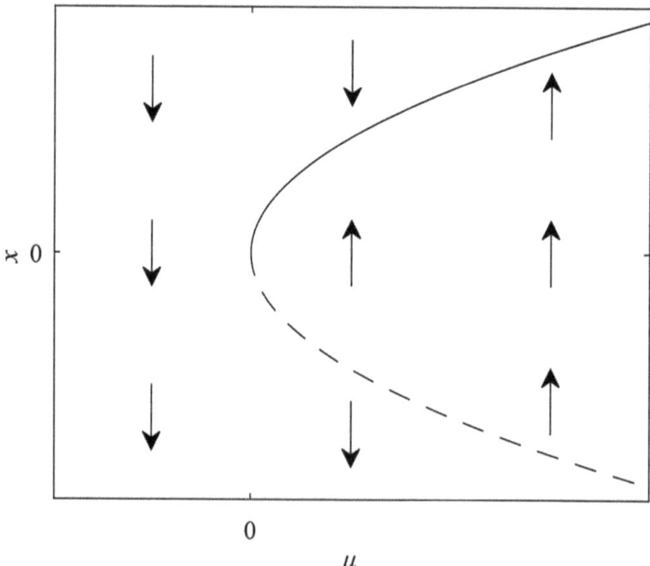

Fig. 6.2 The bifurcation diagram for the saddle–node bifurcation (6.5). The solid line is the stable fixed point and the dashed line is the unstable one. The arrows show the direction of trajectories for a fixed value of μ

a bifurcation to be a saddle–node bifurcation are that $f_{xx} \neq 0$ and $f_\mu \neq 0$ at the bifurcation point.

Example 6.2 Show that (6.2) has two saddle–node bifurcations.

Looking at Fig. 6.1, it appears that (6.2) has saddle–node bifurcations at $\mu = \pm 10$. This can be confirmed by first using Definition 6.1 to find the bifurcation points, and then checking the conditions given above to confirm the type of bifurcation.

In the equation $\dot{x} = f(x, \mu) = 4x^2 + \mu^2 - 100$, a bifurcation occurs if

$$f = 4x^2 + \mu^2 - 100 = 0 \quad \text{and} \quad f_x = 8x = 0.$$

Hence $x = 0$ and $\mu^2 - 100 = 0$, so $\mu = \pm 10$. So there are two bifurcations, at $\mu = 10$, $x = 0$ and at $\mu = -10$, $x = 0$.

One way to show that these bifurcations are saddle–nodes is to expand (6.2) near the bifurcation point. But this has in effect already been done in the derivation of the normal form (6.5). All we need to do is to check the conditions above, $f_{xx} \neq 0$ and $f_\mu \neq 0$. In this case, $f_{xx} = 8 \neq 0$ and $f_\mu = 2\mu = \pm 20 \neq 0$, so indeed both bifurcations are saddle–nodes.

It may seem from the analysis in this section that all bifurcations are saddle–node bifurcations, except in special cases when either f_{xx} or f_μ are zero at the bifurcation point. However, these cases are not exceptional. In the saddle–node bifurcation, there are some parameter values for which no fixed points exist. But in many systems, there is a fixed point that must exist for all parameter values. This situation is considered in Sect. 6.1.2 below. Another common feature is a system with a reflection symmetry $x \leftrightarrow -x$. In this case, $x = 0$ must be a fixed point for all parameter values (see Exercise 3.7) so this fixed point cannot have a saddle–node bifurcation. This case is investigated in Sect. 6.1.3.

6.1.2   Transcritical Bifurcation

Suppose that the first-order differential equation with a parameter (6.1) has the property that $x = 0$ is a fixed point for all values of μ. For example, if the equation is a population model, such as the logistic model (3.9), then $x = 0$ should be a fixed point regardless of the value of any parameter values. Similarly, equations for the rate of change of concentration of a chemical usually have a fixed point where the concentration is zero. Another example is the SIR model (5.25), where there is an infection-free fixed point at $I = 0$ for any value of the parameters and for any value of the other variables S and R. One way of thinking of this feature is that models for populations or chemical concentrations should not allow these quantities to become negative, and having a fixed point at zero ensures this, since trajectories cannot pass through a fixed point.

If there is a fixed point of (6.1) at $x = 0$ for all μ, then $f(0, \mu) = 0$ for all μ, and the fixed point at $x = 0$ cannot have a saddle–node bifurcation. This condition imposes more constraints on the Taylor expansion (6.4). Setting $x = 0$ and imposing $\dot{x} = 0$ for all μ shows that $f_\mu = 0$ and $f_{\mu\mu} = 0$. Then (6.4) simplifies to

$$\dot{x} = \frac{1}{2}x^2 f_{xx} + x\mu f_{x\mu} + O(3).   \tag{6.6}$$

Assuming that $f_{xx} \neq 0$ and $f_{x\mu} \neq 0$, then near the bifurcation, where x and μ are very small, the neglected cubic terms are much smaller than the retained terms. Finally, we can rescale x and μ as in the previous section (see Exercise 6.2) to get

$$\dot{x} = \mu x - x^2.   \tag{6.7}$$

This is the *normal form* for what is known as the *transcritical bifurcation*. The choice of the minus sign for the quadratic term is, as in Sect. 6.1.1, just a convention; we could have chosen to use a plus sign.

The normal form equation (6.7) has two fixed points, $x = 0$, which was the starting assumption, and $x = \mu$. The stability of each fixed point is found from the sign of the derivative of the right-hand side, $\mu - 2x$. The fixed point at $x = 0$ is stable if $\mu < 0$ and unstable if $\mu > 0$, and the fixed point at $x = \mu$ is unstable for $\mu < 0$ and stable for $\mu > 0$. This simple calculation leads to the bifurcation diagram shown in Fig. 6.3. As μ varies, the two fixed points cross over and the stability switches from one branch to the other. For this reason, this type of bifurcation is sometimes known as an *exchange of stabilities*.

In addition to the situation where $x = 0$ is known to be a fixed point for all values of the parameter, this type of bifurcation also occurs in general when two curves representing branches of fixed points cross each other in the bifurcation diagram. Consider a bifurcation point at which $f_\mu = 0$, in which case it cannot be a saddle–node bifurcation. In this case the Taylor series (6.3) truncated after the quadratic terms is

$$\dot{x} = \frac{1}{2}x^2 f_{xx} + x\mu f_{x\mu} + \frac{1}{2}\mu^2 f_{\mu\mu}.   \tag{6.8}$$

Fixed points obey a quadratic equation in x and μ, so if $f_{x\mu}^2 > f_{xx} f_{\mu\mu}$ there are two real roots in which x is proportional to μ, corresponding to two branches of fixed points. In this case the bifurcation equation can be written as

$$\dot{x} = \frac{1}{2}f_{xx}(x - a_1\mu)(x - a_2\mu)   \tag{6.9}$$

for two real numbers a_1 and a_2. Then there are two lines of fixed points, $x = a_1\mu$ and $x = a_2\mu$, that cross at $x = \mu = 0$. It can be checked (see Exercise 6.4) that for any non-zero μ, one of these fixed points is stable and the other is unstable, and the stability switches as μ passes through zero. The bifurcation diagram is a rotated

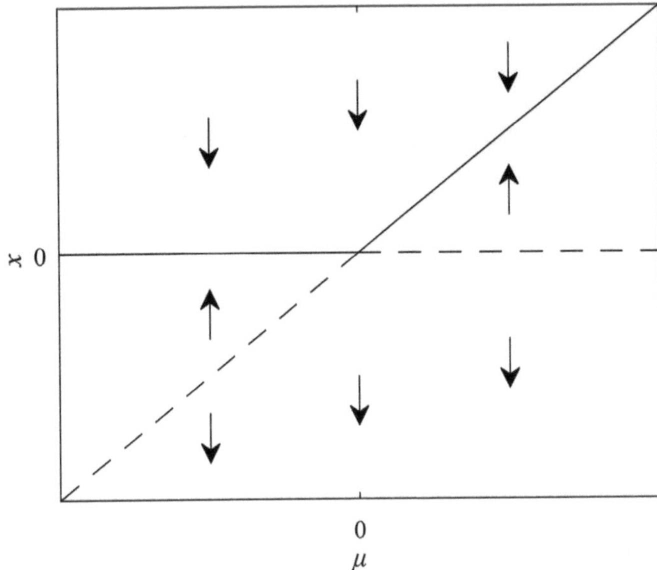

Fig. 6.3 The bifurcation diagram for the transcritical bifurcation (6.7). The solid lines are stable fixed points and dashed lines are unstable ones. The arrows show the direction of trajectories for a fixed value of μ

form of Fig. 6.3 and the bifurcation can be written in the normal form (6.7) by a change of variables and a rescaling. The two branches of fixed points are straight lines, because of the truncation to quadratic terms used in (6.8), but in general, the two branches are curves in the μ, x plane.

In summary, a transcritical bifurcation occurs in (6.1) at a point where $f = f_x = 0$ (the conditions for a bifurcation) and $f_\mu = 0$, $f_{xx} \neq 0$, $f_{x\mu}^2 > f_{xx} f_{\mu\mu}$.

6.1.3 Pitchfork Bifurcation

Some physical systems have a reflection symmetry, $x \leftrightarrow -x$. Examples include the second-order system for a nonlinear pendulum (5.7), a vertical, flexible rod supporting a weight, or the steady flow of water along a straight channel.

This symmetry imposes constraints on the form of (6.1) and on the type of bifurcation that can occur. Suppose that $\dot{x} = f(x, \mu)$ is a model for any system with the symmetry $x \leftrightarrow -x$. Changing the sign of x gives $-\dot{x} = f(-x, \mu)$, so $\dot{x} = -f(-x, \mu)$. The symmetry implies that this must be the same as the original equation, so f must be an odd function of x, $f(x, \mu) = -f(-x, \mu)$. Hence $f(0, \mu) = 0$, so $x = 0$ is a fixed point for all μ. This suggests that a bifurcation in a symmetric system may be transcritical, but it is not, because the fact that f is an odd function of x means that $f_{xx}(0, \mu) = 0$, so one of the conditions for a transcritical bifurcation does not hold.

Now consider the Taylor expansion up to quadratic terms (6.4) near a bifurcation point, with this symmetry constraint. The only term in (6.4) that is odd in x is the $x\mu f_{x\mu}$ term, but with only this linear term in the equation, the solution would simply be exponential growth or decay. To balance this term, and avoid infinite exponential growth, the cubic terms in the Taylor expansion must be included. There are four such terms in general, but only two are odd in x, leading to

$$\dot{x} = x\mu f_{x\mu} + \frac{1}{6}x^3 f_{xxx} + \frac{1}{2}x\mu^2 f_{x\mu\mu} + O(4).$$

Looking at the size of these terms for small x and μ, it is clear that we can drop the last term, as long as $f_{x\mu} \neq 0$, since $x\mu^2 \ll x\mu$. Therefore we can consider the truncated Taylor series in the form

$$\dot{x} = x\mu f_{x\mu} + \frac{1}{6}x^3 f_{xxx}.$$

As in the previous two cases, we now rescale the variables to get a simplified normal form, but in this case something slightly different happens. Setting $x = a\hat{x}$ and $\mu = c\hat{\mu}$ and cancelling a factor of a leads to

$$\dot{\hat{x}} = c\hat{x}\hat{\mu} f_{x\mu} + \frac{1}{6}a^2\hat{x}^3 f_{xxx}.$$

Now we try to choose a and c to remove all the constants and reduce the equation to

$$\dot{x} = \mu x - x^3. \tag{6.10}$$

To do this we choose $c = 1/f_{x\mu}$, which is possible as long as $f_{x\mu} \neq 0$. We also need $a^2 f_{xxx}/6 = -1$, so $a = \sqrt{-6/f_{xxx}}$, but this is only possible if $f_{xxx} < 0$. If $f_{xxx} > 0$, we can choose $a = \sqrt{6/f_{xxx}}$, leading to the normal form

$$\dot{x} = \mu x + x^3. \tag{6.11}$$

Hence there are two different possible forms of the bifurcation. If $f_{xxx} < 0$, the normal form is (6.10), which is known as a *supercritical pitchfork bifurcation*. But if $f_{xxx} > 0$, the normal form is (6.11), called a *subcritical pitchfork bifurcation*. Unlike the two previous bifurcations, there is no change of variables that can translate one form into the other.

In both cases, the fixed point $x = 0$ is stable for $\mu < 0$ and unstable for $\mu > 0$. In (6.10) there are two further fixed points, $x = \pm\sqrt{\mu}$, that exist for $\mu > 0$. The x-derivative of (6.10) at these fixed points is $\mu - 3x^2 = -2\mu$, so these solutions are always stable when they exist. The bifurcation diagram is shown in Fig. 6.4a, which

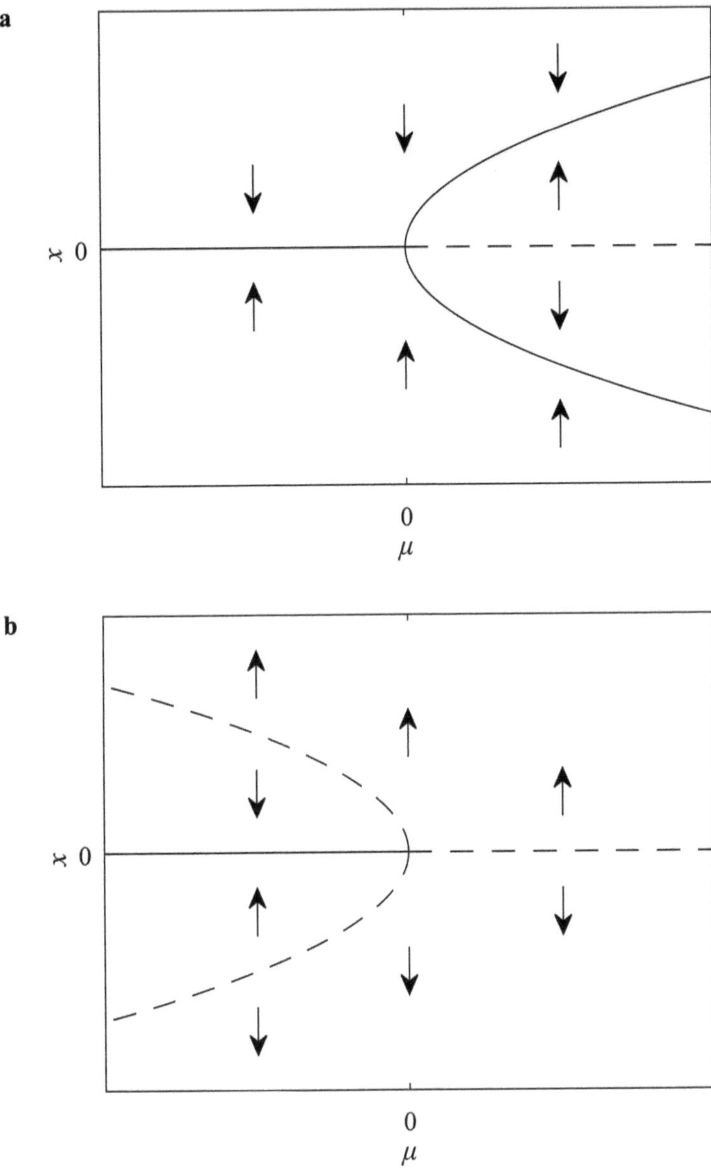

Fig. 6.4 (**a**) Bifurcation diagram for the supercritical pitchfork bifurcation, where as μ increases, the fixed point $x = 0$ becomes unstable and two stable fixed points are created. (**b**) The subcritical case, where the stability of the point $x = 0$ is the same, but two unstable fixed points exist for $\mu < 0$

explains why the bifurcation is called a pitchfork. The term *supercritical* indicates that the additional solutions exist when μ is greater than the critical value $\mu = 0$. In the subcritical case, see Fig. 6.4b, there are fixed points for $\mu < 0$ at $x = \pm\sqrt{-\mu}$, and these are always unstable.

In a supercritical bifurcation, as the parameter μ is increased, the stable large-t solution switches from the state $x = 0$ to one of the two stable states $x = \pm\sqrt{\mu}$. But in the subcritical case, there is no stable state for $\mu > 0$, and according to the normal form (6.11), trajectories tend to infinity. In most practical problems, there is a larger-amplitude stable state that the trajectories approach when $\mu > 0$. See Exercise 6.6 for an example.

6.2 Bifurcations in Higher-Order Differential Equations

In second-order and higher-order systems, there are more types of bifurcation that can occur. But the good news is that the three types of bifurcation derived in the previous section for first-order equations all occur often in higher-order systems. The reason for this is presented in Sect. 6.2.1. These are known as *stationary bifurcations*, since they involve an eigenvalue of a fixed point passing through zero.

A new type of bifurcation called a *Hopf bifurcation* or *oscillatory bifurcation* can occur in systems of order $n > 1$. In this case, the eigenvalues of a fixed point are complex and a fixed point changes from a stable spiral to an unstable spiral as the bifurcation point is crossed, see Sect. 6.2.2. At a Hopf bifurcation, a periodic orbit is created.

Both stationary and oscillatory bifurcations are examples of *local bifurcations*. This means that the stability of a fixed point changes, and this only influences the phase plane in a small, local region near the fixed point that is undergoing the bifurcation.

A very different type of bifurcation is known as a *global bifurcation*, discussed in Sect. 6.2.3. This type of bifurcation, which can only occur in systems of order $n > 1$, does not involve a change to the stability of a fixed point, and cannot be understood by a local Taylor expansion. A global bifurcation usually involves an intersection between a periodic orbit and a fixed point, leading to the disappearance of the periodic orbit or a change in its structure.

In summary, bifurcations in second-order and higher-order systems can be divided into local bifurcations, that only influence a small region of phase space near a fixed point, and global bifurcations, that change the large-scale structure of the phase plane. Local bifurcations may be stationary, if a real eigenvalue passes through zero, or oscillatory, if the real part of a complex eigenvalue goes through zero.

6.2.1   Stationary Bifurcations and Reduction of Order

Definition 6.2 Suppose that $\dot{x} = f(x, \mu)$ is a system of n differential equations including a real parameter μ, so x and f are vectors of length n. The system has a *stationary bifurcation* at $x = x_0$, $\mu = \mu_0$ if $f(x_0, \mu_0) = 0$ and the determinant of the Jacobian matrix J evaluated at (x_0, μ_0) is zero. As in the first-order case, a 'crossing condition' is also required: the sign of the determinant must change as the point (x_0, μ_0) is crossed.

Since the determinant is the product of the eigenvalues, at least one of the eigenvalues of J must be equal to zero at a bifurcation. We will assume that only one of the eigenvalues is zero. This is generally the case, unless the system has a high degree of symmetry, which can lead to the existence of two or more identical eigenvalues. We will also suppose that all the other eigenvalues have negative real part, although this is not essential for the content of this section. So there is a stationary bifurcation when a fixed point crosses from stable to unstable as a parameter varies. At the bifurcation point, the linearised system corresponds to a line of fixed points (see Table 5.1). This line of fixed points is aligned with the eigenvector corresponding to the zero eigenvalue.

In some special cases, such as the SIR model of Sect. 5.1.3, the line of fixed points also occurs in the full nonlinear system, but in general, the line of fixed points is a consequence of the linearisation. In the nonlinear system, the line of fixed points becomes an invariant curve in phase space. Near the bifurcation, trajectories approach this curve exponentially, and then move more slowly along it, effectively reducing the order of the system down to one. This invariant curve is called the *centre manifold*.

Definition 6.3 If $\dot{x} = f(x, \mu)$ has a stationary bifurcation at $x = x_0$, $\mu = \mu_0$, with a single zero eigenvalue, the *centre manifold* at $\mu = \mu_0$ is an invariant curve that passes through the fixed point and is tangent to the eigenvector corresponding to the zero eigenvalue.

Example 6.3 We have already seen an example of a centre manifold, in Example 5.2 and Fig. 5.1, for the system

$$\dot{x} = y - x^2,$$

$$\dot{y} = -y.$$

In this example, the origin is a fixed point with a zero eigenvalue with eigenvector $(1, 0)^T$ pointing along the x axis, and the linear system has a line of fixed points on the x axis, shown in Fig. 5.1a. The centre manifold is therefore an invariant curve that passes through the origin parallel to the line $y = 0$. Since $y = 0$ is an invariant line in the nonlinear system, the centre manifold in this example is simply $y = 0$. The other eigenvalue is -1, so solution trajectories approach the line $y = 0$ exponentially, and then move more slowly along that line, according to $\dot{x} = -x^2$.

If $x > 0$, x decreases to zero as t increases, in a manner proportional to $1/t$, much more slowly than the exponential decay onto the line $y = 0$.

This example illustrates the key point: near a stationary bifurcation, trajectories move rapidly onto the centre manifold, and then move slowly along it. In effect, solutions are squashed down onto this one-dimensional invariant curve, as is apparent in Fig. 5.1b. By finding the centre manifold, we can resolve the neutral stability of the fixed point and determine whether it is stable or unstable.

The Taylor series of the centre manifold can be calculated to any required order, using the fact that it is an invariant curve and its initial direction is known, as shown in the following example.

Example 6.4 Find the first two non-zero terms of the Taylor series of the centre manifold of the origin for the system

$$\dot{x} = 2xy - xy^2, \tag{6.12}$$

$$\dot{y} = -y + x^2, \tag{6.13}$$

and hence determine whether the origin is stable or unstable.

The origin is a fixed point. The Jacobian at the origin and its eigenvalues and eigenvectors are

$$J = \begin{pmatrix} 0 & 0 \\ 0 & -1 \end{pmatrix}, \quad \lambda_1 = 0, v_1 = \begin{pmatrix} 1 \\ 0 \end{pmatrix}, \quad \lambda_2 = -1, v_2 = \begin{pmatrix} 0 \\ 1 \end{pmatrix}.$$

The linear system, $\dot{x} = 0$, $\dot{y} = -y$, is neutrally stable, with a line of fixed points along the x axis, $y = 0$. Whether the origin is stable or unstable depends on the nonlinear terms. From Definition 6.3, the centre manifold must pass through the origin tangent to the line $y = 0$, so its Taylor series has no constant term and no linear term, and can be written as

$$y = ax^2 + bx^3 + cx^4 + O(x^5).$$

Although only two non-zero terms are asked for, we have chosen to include three terms, in case one of them turns out to be zero, which is the case in this example. The centre manifold must be an invariant curve, so on the curve $h(x, y) = y - ax^2 - bx^3 - cx^4 + O(x^5) = 0$, $\dot{h} = 0$ (see Definition 5.2). This condition enables us to find the constants a, b and c, and higher terms in the Taylor series if they are needed. The invariant curve condition gives[1]

$$\dot{y} - (2ax + 3bx^2 + 4cx^3 + O(x^4))\dot{x} = 0,$$

---

[1] Note that differentiating the $O(x^5)$ term gives an $O(x^4)$ term.

and using (6.12) and (6.13) for $\dot{x}$ and $\dot{y}$ leads to

$$-y + x^2 - (2ax + 3bx^2 + 4cx^3 + O(x^4))(2xy - xy^2) = 0. \qquad (6.14)$$

Finally, we substitute in the Taylor series for y, to get an equation involving x only in which we can equate coefficients of powers of x to find a, b and c. Expanding all the terms would give a cumbersome expression, but many of the higher order terms are not needed. Since the centre manifold includes the cx^4 term, it is consistent to keep terms up to this order and write all the higher order terms as $O(x^5)$. In the multiplication of the two brackets in (6.14), we only need to keep the product $(2ax)(2xy) = 4ax^2(ax^2 + O(x^3)) = 4a^2x^4 + O(x^5)$. Then (6.14) becomes

$$-ax^2 - bx^3 - cx^4 + x^2 - 4a^2x^4 + O(x^5) = 0.$$

Comparing the x^2 terms gives $a = 1$, the cubic terms show that $b = 0$, and from the x^4 terms, $-c - 4a^2 = 0$ so $c = -4$. It is no coincidence that $b = 0$. With the benefit of hindsight, we could have realised this in advance and saved some work. The original system has a symmetry $x \leftrightarrow -x$, since (6.12) only includes odd powers of x and (6.13) only has even powers of x, so changing the sign of x leaves both equations unchanged. Hence the centre manifold only has even powers of x, and is

$$y = x^2 - 4x^4 + O(x^6).$$

Near the origin, trajectories decrease exponentially in the y direction onto this curve.
   On the centre manifold, substituting y into (6.12) gives

$$\dot{x} = 2x(x^2 - 4x^4 + O(x^6)) - x(x^2 - 4x^4 + O(x^6))^2 = 2x^3 - 9x^5 + O(x^7).$$

The first term is the dominant one for small x, so the conclusion is that the origin is unstable (because for $\dot{x} = x^3$, solutions move away from the origin). The phase plane is shown in Fig. 6.5, which shows how trajectories near the origin drop rapidly onto the centre manifold curve and then move along it.

Example 6.4 shows how the centre manifold can be found at a bifurcation point, where there is a zero eigenvalue. Then this can be used to obtain a first-order equation on the centre manifold, to find out whether the fixed point is stable or unstable in the direction of the zero eigenvalue.

Now suppose that a system with a parameter has a stationary bifurcation at $\mu = \mu_0$, and that the parameter is near the bifurcation point, so $\mu = \mu_0 + \varepsilon$, where ε is small. In that case, we can find an *extended centre manifold* (ECM), via a Taylor series expansion for y in two variables, x and ε. Then, in a similar way, on this curve, the system is reduced to a first-order equation for $\dot{x}$ in terms of x and the parameter ε. This equation can then be identified as one of the standard bifurcations of Sect. 6.1, showing that these three simple bifurcations do frequently occur in higher order

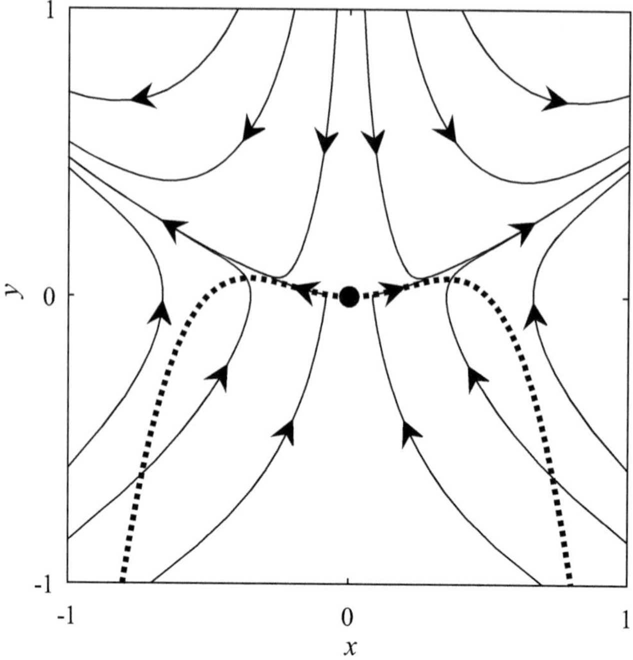

Fig. 6.5 Phase portrait for Example 6.4. The dotted line is the first two terms in the Taylor series of the centre manifold, $y = x^2 - 4x^4$. Near the origin, solutions move rapidly onto this curve and then move slowly along it, away from the origin

systems. This is best illustrated by specific examples. The two following examples are based on modifications of systems already studied in Chap. 5.

Example 6.5 Consider the Lotka–Volterra model of Example 5.8 with a variable parameter for the growth rate of the x species,

$$\dot{x} = \mu x - x^2 - xy,$$
$$\dot{y} = 4y - 2y^2 - xy.$$

Show that there is a fixed point at $(0, 2)$ for all μ and find the value of μ where this fixed point has a bifurcation. Find the extended centre manifold to linear order, and hence identify the type of bifurcation.

At a fixed point, $x(\mu - x - y) = 0$ and $y(4 - 2y - x) = 0$, so as in Example 5.8 there are in general four fixed points, at $(0, 0)$, $(0, 2)$, $(\mu, 0)$ and $(2\mu - 4, 4 - \mu)$. So $(0, 2)$ is a fixed point for all μ, while the location of the other two non-zero fixed points depends on μ. The Jacobian is

$$J = \begin{pmatrix} \mu - 2x - y & -x \\ -y & 4 - 4y - x \end{pmatrix} = \begin{pmatrix} \mu - 2 & 0 \\ -2 & -4 \end{pmatrix} \quad \text{at the fixed point } (0, 2).$$

The eigenvalues are $\mu - 2$ and -4, so the fixed point is stable for $\mu < 2$ and unstable for $\mu > 2$. At $\mu = 2$ there is a zero eigenvalue, and hence a bifurcation, according to Definition 6.2.

The quick way to see what type of bifurcation occurs at $\mu = 2$ is to note that the fixed point $(0, 2)$ exists for all μ, which rules out a saddle–node bifurcation. Also, the fourth fixed point, $(2\mu - 4, 4 - \mu)$, passes through $(0, 2)$ when $\mu = 2$ and enters the (biologically meaningless) region $x < 0$ for $\mu < 2$. Since two fixed points cross at the bifurcation point, it seems very likely that there is a transcritical bifurcation. The ECM can be used to confirm this by actually deriving the normal form (6.7).

To find the extended centre manifold, we set $\mu = 2 + \varepsilon$ and find the equation of an invariant curve passing through the fixed point, tangent to the zero eigenvector, as a power series for $y(x, \varepsilon)$. At $\mu = 2$, the eigenvector of the zero eigenvalue is $(2, -1)^T$, so the ECM is tangent to the straight line passing through $(0, 2)$ with slope $-1/2$, which is the line $y = 2 - x/2$, when $\varepsilon = 0$. The ECM is therefore a power series in x and ε that can be written as

$$y = 2 - x/2 + a\varepsilon + O(2),$$

where $O(2)$ represents the quadratic and higher order terms in x and ε, which are not needed to identify the bifurcation. The curve must pass through the fixed point $y = 2$ when $x = 0$ for all values of ε, so $a = 0$.

Now we substitute this formula for y into the $\dot{x}$ equation:

$$\dot{x} = (2 + \varepsilon)x - x^2 - x\left(2 - \frac{x}{2} + O(2)\right)$$

$$= \varepsilon x - \frac{x^2}{2} + O(3),$$

which is the normal form for a transcritical bifurcation (6.7), after rescaling x by a factor of 2.

Example 6.6 Consider the damped, nonlinear pendulum from Sect. 5.1.2, with the addition of a constant force acting to rotate the pendulum in a particular direction. The equations are (5.8), (5.9) with an additional constant forcing term, $\mu \geq 0$:

$$\dot{x} = y, \tag{6.15}$$

$$\dot{y} = -\beta y - k \sin x + \mu, \tag{6.16}$$

where $k = g/L$ has been introduced to simplify the notation. This system also models an electronic device known as a Josephson Junction. Suppose that the parameters β and k are fixed, but μ is varied as the bifurcation parameter. Fixed points occur where $y = 0$ and $k \sin x = \mu$. For $\mu = 0$ there is a saddle at $x = \pi$ and a stable point at $x = 0$ that may be a stable node or a stable spiral (see Example 5.9 and Exercise 5.11). If $\mu > 0$, there are still two fixed points in $0 \leq x < 2\pi$ as long as $\mu < k$. As μ approaches k, $\sin x$ approaches 1 at the fixed points, so both fixed points are close to $x = \pi/2$. If $\mu = k$ there is only one fixed point, $x = \pi/2$, and for $\mu > k$, there are no fixed points. This suggests that there is a saddle–node bifurcation at $\mu = k$.

Following a similar process to the previous example confirms this mathematically. Some of the steps will be omitted here and left as an exercise. Using Definition 6.2, the determinant of the Jacobian is $k \cos x$, so there is a bifurcation at $x = \pi/2$, and the fixed point condition gives $\mu = k$, $y = 0$. At the bifurcation, the eigenvector of the zero eigenvalue is $(1, 0)^T$, so the centre manifold is horizontal at this point. Therefore the CM is a curve passing through $(\pi/2, 0)$ with zero slope,

$$ y = a(x - \pi/2)^2 + O(x - \pi/2)^3 = az^2 + O(z^3), $$

where $z = x - \pi/2$ is assumed to be small, so we are close to the fixed point. The invariant curve condition $\dot{y} - 2az\dot{z} = 0$ can be used to find a, leading to

$$ -\beta az^2 - k \cos z + k + O(z^3) = 0, $$

and after expanding $\cos z$ for small z (see Sect. A.2.3), the z^2 terms give $a = k/2\beta$, so the CM is $y = kz^2/2\beta + O(z^3)$.

Now to derive the type of the bifurcation, set $\mu = k + \varepsilon$ and find the ECM by writing y as a series in ε and z. Unlike the previous example, the location of the fixed point depends on ε. The fixed point condition $\mu = k \sin x$ gives $k + \varepsilon = k \cos z$, and using the series for $\cos z$ again shows that the fixed points occur at $\varepsilon = -kz^2/2 + O(z^4)$. So fixed points only occur if $\varepsilon < 0$. This confirms our suspicion that we have a saddle–node bifurcation, but our aim is to actually derive (6.5) to show that this is correct. Since the ECM must pass through the fixed points, the formula for y must be zero when $z^2 = -2\varepsilon/k$. The ECM is then

$$ y = \frac{k}{2\beta}\left(z^2 + \frac{2\varepsilon}{k} + O(z^3)\right) = \frac{\varepsilon}{\beta} + \frac{k}{2\beta}z^2 + O(z^3). \tag{6.17} $$

Finally, to identify the bifurcation, we substitute this into the equation $\dot{x} = y$:

$$ \dot{z} = \frac{\varepsilon}{\beta} + \frac{k}{2\beta}z^2 + O(z^3), $$

which after rescaling is the normal form for a saddle–node bifurcation (6.5).

Figure 6.6 shows the phase portrait for parameters $\beta = 1.5$, $k = 1$, $\mu = 0.9$, fairly near the saddle–node bifurcation at $\mu = 1$. As in the previous example, a typical trajectory moves rapidly onto the centre manifold curve and then moves slowly along it. The two fixed points are a saddle point and a stable node.

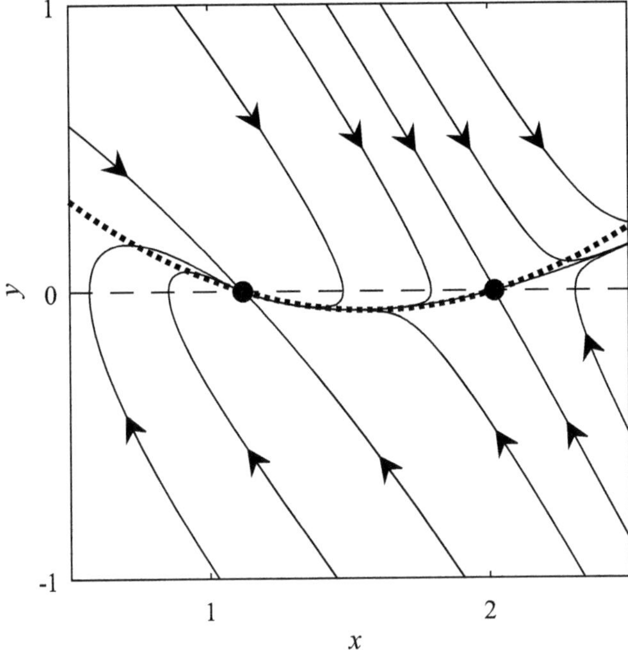

Fig. 6.6 Phase plane for the damped, forced pendulum, Example 6.6, with $\beta = 1.5, k = 1, \mu = 0.9$. The dashed line is the x axis, which is the y-nullcline. The dotted line is the approximation of the extended centre manifold, (6.17)

Remark 6.2 In each of the previous two examples, it was possible to deduce the type of bifurcation just by looking at the existence and stability of fixed points. The reduction to a first-order normal form using the extended centre manifold provides useful confirmation, and helps us to understand why the bifurcations of first-order differential equations frequently occur in higher-order systems, but it is not essential to do this for every problem. The following example illustrates this approach.

Example 6.7 By considering the fixed points and their stability, determine the type of bifurcation that occurs at $\mu = 0$ in the system

$$\dot{x} = \mu x + x^3 + xy, \tag{6.18}$$

$$\dot{y} = -y - 2x^2. \tag{6.19}$$

Clearly the origin is a fixed point for all μ, so we can deduce immediately that the origin does not have a saddle–node bifurcation. The Jacobian at the origin is diagonal, with eigenvalues μ and -1, so the origin is stable for $\mu < 0$ and unstable for $\mu > 0$, with a bifurcation at $\mu = 0$. If there are any other fixed points, then from (6.19), $y = -2x^2$, so using (6.18), $\mu x + x^3 - 2x^3 = 0$, hence either $x = 0$ or

$x = \pm\sqrt{\mu}$. At these two fixed points, $y = -2\mu$, and they only exist for $\mu > 0$. By finding the eigenvalues of the Jacobian it can be shown that these two fixed points are stable for small, positive μ. For $\mu < 0$ there is only one fixed point, at the origin, and it is stable. For $\mu > 0$, the origin is unstable and there are two stable fixed points at $x = \pm\sqrt{\mu}$. From this pattern of the fixed points and their stability it is easy to see that this system has a supercritical pitchfork bifurcation at $\mu = 0$.

Another way to deduce the type of bifurcation in this example is to consider the symmetry of the system. In (6.18) every term is an odd function of x, while in (6.19) every term is even in x. So if we reverse the sign of x, both equations are unchanged. Since the equations have the symmetry $x \leftrightarrow -x$, the bifurcation from the origin is a pitchfork. The symmetry does not tell us whether the pitchfork is supercritical or subcritical. This has to be deduced by finding the fixed points. In fact it is not necessary to investigate their stability as described in the previous paragraph, because we know that the non-zero fixed points exist when the zero state is unstable, so the pitchfork must be supercritical.

Remark 6.3 The method described in this section for second-order systems can be extended to higher order. Provided that only one eigenvalue is zero at a bifurcation, we can find a one-dimensional centre manifold near the bifurcation point, and then look at the behaviour on this invariant curve to find the type of bifurcation.

In systems with a greater level of symmetry than that of Example 6.7, the symmetry can force more than one eigenvalue to be zero. In this case, the dimension of the centre manifold is greater than one, and the types of bifurcations that can occur are more complicated. This provides another example of links between apparently unrelated topics in mathematics. Group theory and representation theory are needed for a proper understanding of the effects of symmetry on differential equations and bifurcations.

Another situation where the centre manifold is two-dimensional is when a fixed point has a complex conjugate pair of purely imaginary eigenvalues. This case is considered in the following section, and an example of the computation of the centre manifold will be seen in Chap. 8.

6.2.2   Hopf Bifurcation

Up to this point we have considered *stationary bifurcations*, in which an eigenvalue of a fixed point passes through zero and the stability of the fixed point changes. In systems of order two and higher, a fixed point can also change stability if it has a complex conjugate pair of eigenvalues and the real part of these eigenvalues passes through zero. Consider the linear system

$$\begin{pmatrix} \dot{x} \\ \dot{y} \end{pmatrix} = \begin{pmatrix} \mu & -\omega \\ \omega & \mu \end{pmatrix} \begin{pmatrix} x \\ y \end{pmatrix}, \tag{6.20}$$

where $\omega \neq 0$ is fixed and μ is regarded as a variable parameter. The eigenvalues of the fixed point at the origin are $\lambda = \mu \pm i\omega$, so the linear system is a stable spiral for $\mu < 0$, a centre for $\mu = 0$, and an unstable spiral for $\mu > 0$. The solutions for $x(t)$ and $y(t)$ are of the form $e^{\mu t}(A \cos \omega t + B \sin \omega t)$. The change in the behaviour as μ passes through zero is known as an *oscillatory bifurcation* or *Hopf bifurcation*.

When a Hopf bifurcation occurs in a system of order greater than two, a centre manifold can be defined, which is two-dimensional. The nonlinear behaviour can be analysed using Taylor series in a similar way to the bifurcations derived in Sect. 6.1, but the calculations are extremely cumbersome.

The end result is that a Hopf bifurcation can be reduced to the normal form

$$\dot{x} = \mu x - \omega y - \alpha x(x^2 + y^2) + \beta y(x^2 + y^2), \tag{6.21}$$

$$\dot{y} = \omega x + \mu y - \alpha y(x^2 + y^2) - \beta x(x^2 + y^2), \tag{6.22}$$

for some constants α and β. This normal form can be written more neatly in two alternative forms. The first form uses polar coordinates, (r, θ), as in Example 5.11, which looked at the special case where $\mu = \omega = \alpha = 1$ and $\beta = 0$. Following the method given there leads to

$$\dot{r} = \mu r - \alpha r^3, \tag{6.23}$$

$$\dot{\theta} = \omega - \beta r^2. \tag{6.24}$$

The Hopf bifurcation normal form can be written as a single differential equation by defining a complex number $z = x + iy = re^{i\theta}$. Then

$$\dot{z} = (\mu + i\omega)z - (\alpha + i\beta)|z|^2 z. \tag{6.25}$$

It is clear from (6.23) that a Hopf bifurcation looks very like a pitchfork bifurcation. There is a fixed point at $r = 0$ that is stable for $\mu < 0$ and unstable for $\mu > 0$. If $\alpha > 0$, (6.23) has a stable fixed point at $r = \sqrt{\mu/\alpha}$ that exists for $\mu > 0$. But this is not a fixed point of the whole system, because in general $\dot{\theta} \neq 0$, so it represents a stable periodic orbit, similar to that shown in Fig. 5.8. This case is called a *supercritical Hopf bifurcation*. For $\mu > 0$, solutions that start near the origin spiral outwards and approach the stable periodic orbit. But if $\alpha < 0$, the fixed point of (6.23) at $r = \sqrt{\mu/\alpha}$ exists for $\mu < 0$ and is always unstable, corresponding to an unstable periodic orbit. This is a *subcritical Hopf bifurcation*.

The bifurcation diagram for a Hopf bifurcation is a three-dimensional picture, plotting x and y as a function of μ, see Fig. 6.7, which shows the supercritical case. It can be simplified by just plotting a cross-section, in which case the bifurcation

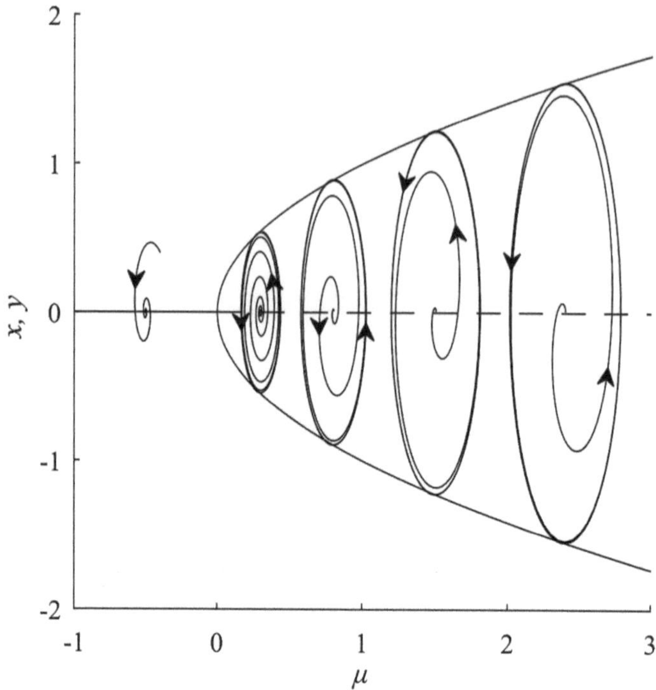

Fig. 6.7 Bifurcation diagram for the supercritical Hopf bifurcation. For $\mu < 0$, trajectories spiral in towards the origin. For $\mu > 0$, the origin is an unstable spiral and solutions approach the stable periodic orbit. The trajectories shown here show the phase plane behaviour for different values of μ, but these would not normally be drawn on a bifurcation diagram

diagram looks exactly like the bifurcation diagram of a pitchfork bifurcation (Fig. 6.4), except that it must be remembered that the solid or dashed line with $r \neq 0$ represents a periodic orbit, not a fixed point.

To find a Hopf bifurcation, we just need to find the eigenvalues of the Jacobian at a fixed point and see when they are purely imaginary. It is much harder to find out whether the Hopf bifurcation is supercritical or subcritical. There is a horrendous formula in terms of second and third derivatives of a second-order system that can be used to calculate the value of α, or it can be worked out in a specific case, see Example 6.8 below. The parameter β is less important. It represents the speeding up or slowing down of the rotation rate of the spiralling solutions as r changes.

Example 6.8 Show that the equation

$$\ddot{x} + (x^2 - \mu)\dot{x} + x = 0 \tag{6.26}$$

has a Hopf bifurcation at $\mu = 0$, and investigate whether the bifurcation is supercritical or subcritical.

First write the equation as a second order system,

$$\dot{x} = y, \qquad \dot{y} = -x + \mu y - x^2 y.$$

The only fixed point is $x = y = 0$. The Jacobian at the origin is

$$J = \begin{pmatrix} 0 & 1 \\ -1 & \mu \end{pmatrix}$$

so the eigenvalues obey $\lambda^2 - \mu\lambda + 1 = 0$. To seek a Hopf bifurcation, set $\lambda = i\omega$, with $\omega \neq 0$, so that the eigenvalue is imaginary. From the real part of the eigenvalue equation we get $\omega^2 = 1$, and the imaginary part gives $\mu\omega = 0$, so there is a Hopf bifurcation at $\mu = 0$. The origin is a centre at $\mu = 0$, a stable spiral for $-2 < \mu < 0$ and an unstable spiral for $0 < \mu < 2$.

To see whether the Hopf bifurcation is supercritical or subcritical, suppose that μ is small, so we are near the bifurcation point. According to the Hopf bifurcation normal form, there should be a periodic orbit in which x is also small, and the relative magnitude of μ and x should be $x = O(\mu^{1/2})$. In (6.26) the $O(\mu^{1/2})$ terms are $\ddot{x} + x = 0$, which has the general solution $x = A \sin(t - t_0)$, for a constant A that we would like to find, and a constant t_0 that we can set to 0 by choosing our time origin. To find the value of A, a useful method is to multiply (6.26) through by $\dot{x}$ and then integrate over one period T of the periodic orbit. Since

$$\int_0^T \dot{x}\ddot{x}\, dt = \int_0^T \frac{1}{2}\frac{d(\dot{x}^2)}{dt}\, dt = \left[\frac{\dot{x}^2}{2}\right]_0^T = 0,$$

the $\ddot{x}$ term makes no contribution to the integral, and nor does the x term, for the same reason. This leaves

$$\int_0^T (x^2 - \mu)\dot{x}^2\, dt = 0,$$

which is often referred to as a *solvability condition*. Substituting in the small-amplitude solution $x = A \sin t$ gives

$$\int_0^{2\pi} (A^2 \sin^2 t - \mu)A^2 \cos^2 t\, dt = 0.$$

This integral can be evaluated using the trig rules $2 \sin t \cos t = \sin 2t$ and $2 \cos^2 t = 1 + \cos 2t$, leading to the result

$$\frac{A^2}{4} - \mu = 0.$$

Hence the periodic orbit exists for $\mu > 0$, when the origin is an unstable spiral, so the Hopf bifurcation must be supercritical and the periodic orbit is stable. The amplitude of the periodic orbit near the bifurcation is $A = \pm 2\sqrt{\mu}$, so a periodic solution of (6.26) for small positive μ is $x \approx 2\sqrt{\mu}\sin t$. In fact this is the first term of an expansion of the periodic solution in odd powers of $\mu^{1/2}$, and the frequency of the oscillation also depends on μ, so $x = 2\mu^{1/2}\sin(1 + O(\mu))t + O(\mu^{3/2})$.

6.2.3   Global Bifurcations

The stationary and oscillatory bifurcations discussed in Sects. 6.1 and 6.2.2 are all *local bifurcations*. A local bifurcation involves the change in stability of a particular fixed point of the system, and the bifurcation only influences the phase space near that fixed point. For example, at the supercritical pitchfork bifurcation (6.10), a stable fixed point at $x = 0$ becomes unstable and two stable fixed points are created, but the qualitative behaviour for large x does not change, see Fig. 6.4a.

At a *global bifurcation*, there is no change to the stability of a fixed point, but the phase portrait changes as a parameter varies. This occurs when a periodic orbit crosses through one or more saddle points. After the bifurcation, the periodic orbit either no longer exists, or may have a different form.

The simplest type of global bifurcation is the *homoclinic bifurcation*, where a periodic orbit collides with a saddle and disappears. This bifurcation is illustrated in the three diagrams in Fig. 6.8. In each figure, there are two fixed points; one of these is an unstable spiral and the other is a saddle. Before the bifurcation occurs, the phase plane picture is as shown in Fig. 6.8a. There is a stable periodic orbit, which encloses the unstable spiral. Lurking just outside the periodic orbit, there is a saddle. As the bifurcation parameter is increased, the periodic orbit grows and gets closer to the saddle, and because the trajectory goes near the fixed point, it slows down, so the period of the orbit gets longer. At the bifurcation point, the orbit meets the saddle, so in fact there is no periodic orbit. In its place there is a *homoclinic connection*, which is a trajectory $(x(t), y(t))$ that approaches the saddle as $t \to -\infty$ and as $t \to \infty$ (Fig. 6.8b). Beyond the bifurcation point, there is no periodic orbit, see Fig. 6.8c. The phase plane diagrams are for the system investigated in the following example.

Example 6.9 Consider the second-order differential equation

$$\ddot{x} + \mu\dot{x} - x + x^2 - \varepsilon x\dot{x} = 0, \tag{6.27}$$

equivalent to the system

$$\dot{x} = y, \quad \dot{y} = x - \mu y - x^2 + \varepsilon xy. \tag{6.28}$$

Here, $\varepsilon > 0$ is a small parameter regarded as fixed, and $\mu \geq 0$, also assumed to be small, is the bifurcation parameter that will be varied.

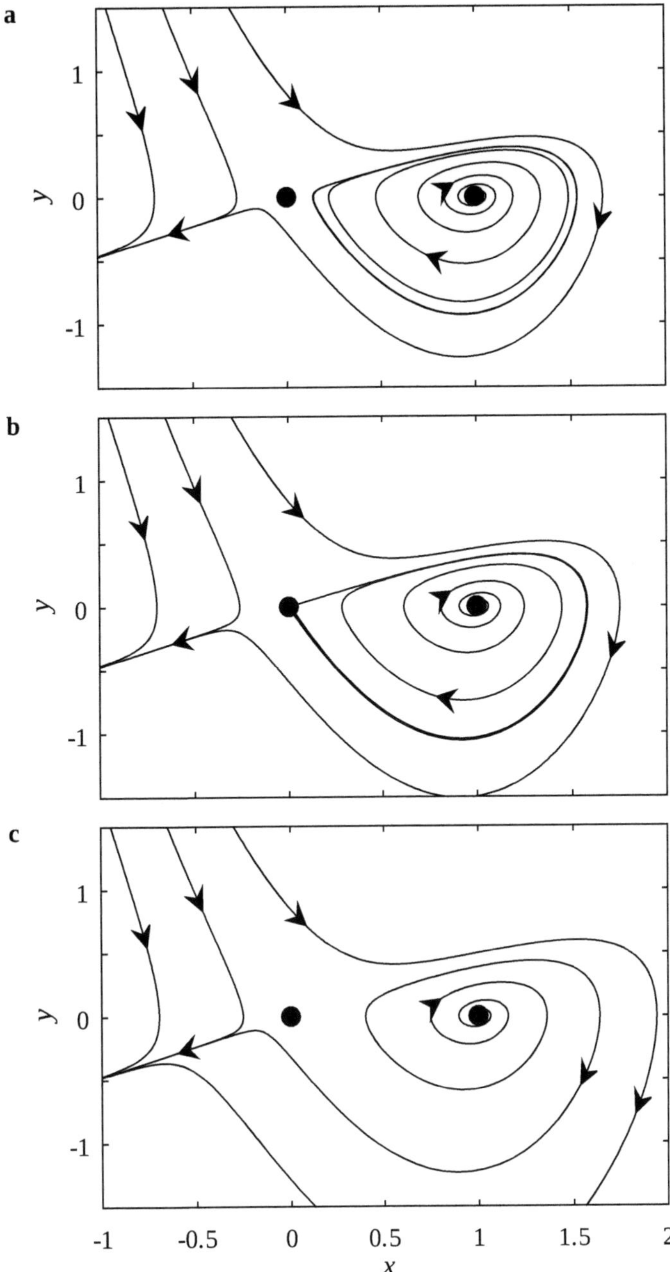

Fig. 6.8 Phase portraits for (6.27), for $\varepsilon = 2$. (**a**) Stable periodic orbit, $\mu = 1.8$, (**b**) homoclinic trajectory, $\mu \approx 1.7705$, (**c**) no periodic orbit, $\mu = 1.72$

The first step is to find the fixed points and investigate their stability. The fixed points are $(0, 0)$ and $(1, 0)$. The Jacobians at a general point and evaluated at the fixed points are

$$
J = \begin{pmatrix} 0 & 1 \\ 1 - 2x + \varepsilon y & -\mu + \varepsilon x \end{pmatrix}, \quad J_{0,0} = \begin{pmatrix} 0 & 1 \\ 1 & -\mu \end{pmatrix}, \quad J_{1,0} = \begin{pmatrix} 0 & 1 \\ -1 & -\mu + \varepsilon \end{pmatrix}.
$$

The fixed point at the origin is always a saddle point, since $|J_{0,0}| = -1$. For the other fixed point, $|J_{1,0}| = 1$ and $T(J_{1,0}) = -\mu + \varepsilon$. Since μ and ε are both assumed to be small, this fixed point has complex eigenvalues. It is a stable spiral if $\mu > \varepsilon$ and an unstable spiral if $\mu < \varepsilon$. For $\mu = \varepsilon$ the fixed point is a centre, with eigenvalues $\pm \mathrm{i}$. At this value of μ there is a Hopf bifurcation, so a periodic orbit is created.

It can be shown that this Hopf bifurcation is supercritical, although the calculation is more time-consuming than for (6.26). This means that there is a stable periodic orbit when the fixed point is an unstable spiral, which is for $\mu < \varepsilon$. However, this statement is only valid near the bifurcation, when μ is close to ε. As μ decreases, the periodic orbit ceases to exist, as shown in Fig. 6.8.

In fact it can be seen that there is no periodic orbit for $\mu = 0$, using the divergence test and the index test from Sects. 5.8.1 and 5.8.3. The divergence of the flow field (6.28) is $-\mu + \varepsilon x$, which is εx when $\mu = 0$. So there cannot be a periodic orbit that remains entirely in the region $x > 0$, by the divergence test, since in that region the divergence is single-signed everywhere. Now by the index test, a periodic orbit must enclose the unstable spiral at $(1, 0)$, so part of it is certainly in $x > 0$. If it crosses into the region where $x < 0$, it must do so in the lower half of the phase plane $y < 0$, since that is where $\dot{x} < 0$, and must return to $x > 0$ in the upper half plane $y > 0$. But then it would have to enclose the saddle point at the origin, which cannot occur, by the index test, so there can be no periodic orbit when $\mu = 0$.

As μ decreases from the Hopf bifurcation at $\mu = \varepsilon$, the periodic orbit must cease to exist at some positive value of μ. This happens at a homoclinic bifurcation, as shown in Fig. 6.8b.

The location of the homoclinic bifurcation cannot be found exactly. But it can be found approximately, with the assumption that ε is small, using the *Melnikov method*. This calculation is quite involved, but it is worthwhile, as the approximation is good even if ε is not very small, and also because it makes use of one of the techniques studied in Chap. 5.

In the limit $\mu \to 0$ and $\varepsilon \to 0$, (6.28) becomes

$$
\dot{x} = y, \quad \dot{y} = x - x^2. \tag{6.29}
$$

This system is Hamiltonian, with the Hamiltonian function

$$
H(x, y) = \frac{y^2}{2} - \frac{x^2}{2} + \frac{x^3}{3}, \tag{6.30}
$$

plus an arbitrary constant that we can set to 0. Now recall from Sect. 5.5 that in a Hamiltonian system, fixed points can only be centres or saddles, and trajectories are the contours of H, the curves $H =$ constant. The fixed points are $(0, 0)$ and $(1, 0)$, as in the full system (6.28), and these are a saddle and a centre respectively. Now consider a trajectory that passes through the saddle at the origin. Since $H(0, 0) = 0$, this trajectory is just the curve $H(x, y) = 0$, so

$$y^2 = x^2 - 2x^3/3. \tag{6.31}$$

This curve passes through the origin twice, tangent to the lines $y = \pm x$, which are the eigenvectors of the origin. It also crosses the line $y = 0$ at $x = 3/2$. So $H(x, y) = 0$ is a homoclinic trajectory, that starts from the origin, crosses the x axis at $x = 3/2$ and then returns to the origin as $t \to \infty$. Since $3/2 > 1$, the homoclinic curve encloses the centre at $(1, 0)$, see Fig. 6.9.

Now for the clever and slightly tricky step. Suppose that we re-introduce the two small terms from (6.28) that were dropped to obtain the Hamiltonian system. Then the Hamiltonian structure is lost, and the centre will become either a stable or unstable spiral, and the homoclinic curve will in general no longer be homoclinic. But when μ and ε are small and related in a particular way, that we can calculate, there is still a homoclinic connection.

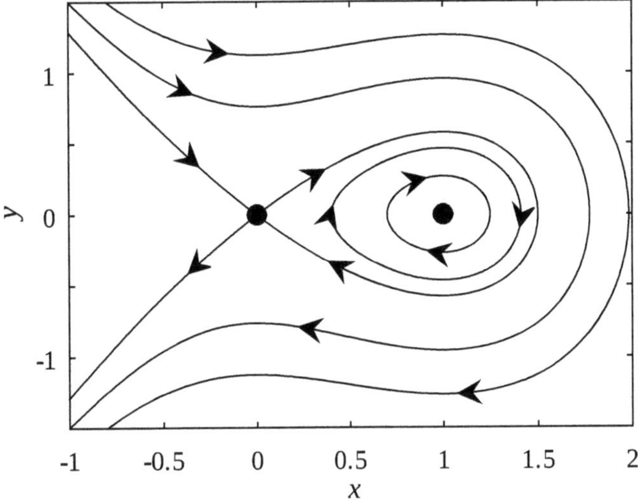

Fig. 6.9 Phase portrait of the Hamiltonian system (6.29). There is a homoclinic trajectory, $H = 0$, connecting the origin to itself, crossing the x axis at $x = 3/2$

Even though (6.28) is not quite Hamiltonian, we can still consider the same function $H(x, y)$ defined in (6.30). If the homoclinic curve exists, linking the saddle point to itself, then on this trajectory, H approaches the same value as $t \to \pm\infty$, so

$$\int_{-\infty}^{\infty} \dot{H}\, dt = [H]_{-\infty}^{\infty} = 0.$$

This condition can be used to find the relationship between μ and ε at the global bifurcation. This integral is

$$\int_{-\infty}^{\infty} H_x \dot{x} + H_y \dot{y}\, dt = \int_{-\infty}^{\infty} (-x + x^2)y + y(x - \mu y - x^2 + \varepsilon xy)\, dt = 0.$$

Because the system is almost Hamiltonian, most of the terms in the integral cancel out, leaving only the contribution from the two small, non-Hamiltonian terms,

$$\int_{-\infty}^{\infty} y(-\mu y + \varepsilon xy)\, dt = 0.$$

In most textbooks this is approached by first explicitly finding $x(t)$ and then doing the integral, but the integral can be evaluated more easily by changing variable from t to x. Since $y = \dot{x}$, this gives

$$\int -\mu y + \varepsilon xy\, dx = 0.$$

This integral is evaluated around the homoclinic curve, consisting of the upper half where $y > 0$ and x increases, followed by the lower half where $y < 0$ and x decreases. These two halves double up, so

$$\int_0^{3/2} (\varepsilon x - \mu) y\, dx = 0,$$

where the limits come from the x range of the homoclinic orbit in the Hamiltonian system found above. Substituting for y using (6.31) (which is an approximation, since in the non-Hamiltonian system, y will be slightly perturbed from the solution (6.31)) gives

$$\int_0^{3/2} (\varepsilon x - \mu) x (1 - 2x/3)^{1/2}\, dx = 0.$$

This integral can be evaluated by making the substitution $u = 1 - 2x/3$, so $x = 3(1-u)/2$ and $dx = -3\,du/2$:

$$\int_1^0 \left(\frac{3}{2}\varepsilon(1-u) - \mu\right)\frac{3}{2}(1-u)u^{1/2}\left(-\frac{3}{2}\right)du = 0.$$

Cancelling numerical factors and collecting powers of u, this simplifies to

$$\int_0^1 (3\varepsilon - 2\mu)u^{1/2} + (2\mu - 6\varepsilon)u^{3/2} + 3\varepsilon u^{5/2}\,du = 0.$$

These integrals can now be evaluated directly, giving

$$(3\varepsilon - 2\mu)\frac{2}{3} + (2\mu - 6\varepsilon)\frac{2}{5} + 3\varepsilon\frac{2}{7} = 0 \quad \Rightarrow \quad \mu = \frac{6}{7}\varepsilon.$$

This formula gives the approximate location at which the homoclinic bifurcation occurs and the periodic orbit is destroyed. To be slightly more precise, the global bifurcation occurs on a curve in the μ, ε plane that is tangent to $\mu = 6\varepsilon/7$ as $\varepsilon \to 0$. The orbit does not last long—it only exists in the small region $6\varepsilon/7 < \mu < \varepsilon$. For $\varepsilon = 2$ this formula suggests that the global bifurcation occurs at $\mu = 12/7 \approx 1.714$. Figure 6.8 shows that this is quite a good approximation, even though $\varepsilon = 2$ is not a very small number.

There are other types of global bifurcation. One of the most elegant is the *gluing* bifurcation, that occurs in the equation

$$\ddot{x} + \mu\dot{x} - x + x^3 - \varepsilon x^2\dot{x} = 0. \tag{6.32}$$

This equation is very similar to (6.27), except that the nonlinear terms are now cubic. This means that (6.32) has a symmetry $x \leftrightarrow -x$, which after defining $y = \dot{x}$ corresponds to a symmetry of rotation through an angle π in the phase plane. There are three fixed points: the origin is a saddle point, and $x = \pm 1$, $y = 0$ are unstable spirals for $\mu < \varepsilon$ if μ and ε are both small.

For $\varepsilon = 1.5$, $\mu = 1.29$, there are two stable periodic orbits, related by the symmetry, see Fig. 6.10a. As μ decreases, these orbits grow until they intersect the saddle at the origin in a global bifurcation, forming a figure-eight or ∞ shape (Fig. 6.10b). After this bifurcation, the two periodic orbits have glued together to form a single large periodic orbit that encloses all three fixed points (Fig. 6.10c).

The Melnikov method can be applied to (6.32) as in Example 6.9. The calculation is very similar, see Exercise 6.16. The result is that for small ε, the gluing bifurcation occurs at approximately $\mu = 4\varepsilon/5$. Again, this agrees fairly well with the phase planes shown in Fig. 6.10, even though ε and μ are not small.

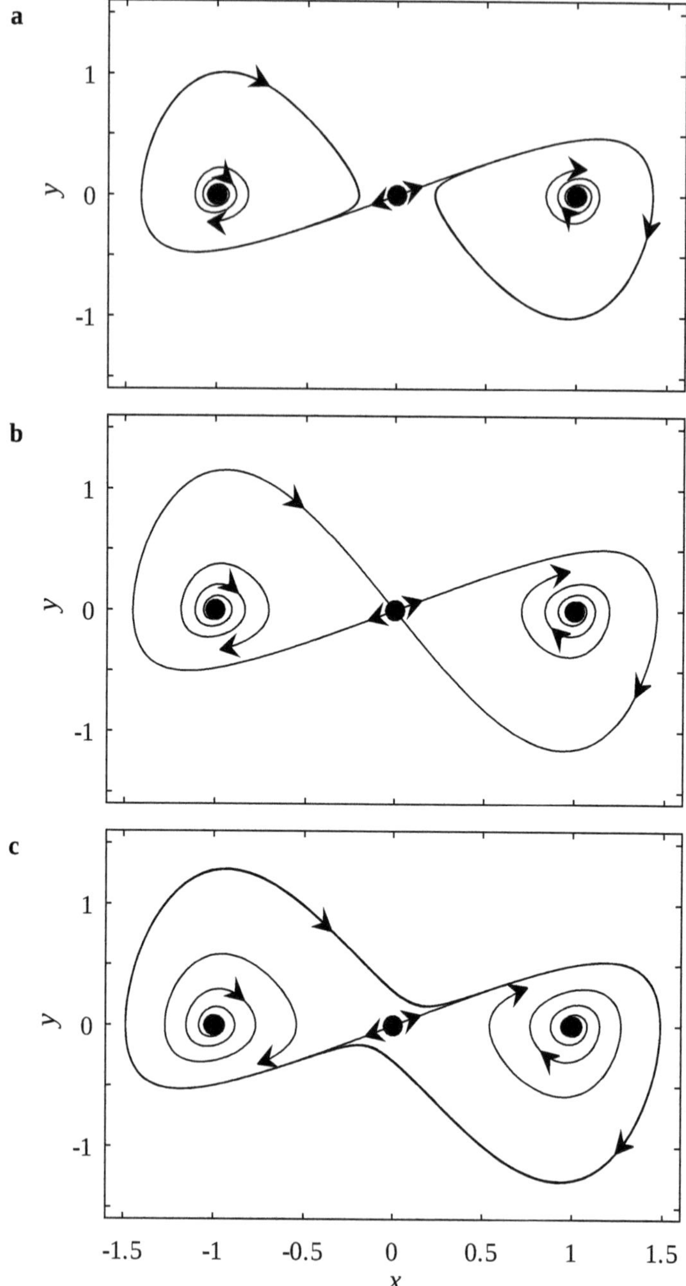

Fig. 6.10 Phase portraits for (6.32), for $\varepsilon = 1.5$. (**a**) $\mu = 1.29$, two periodic orbits. (**b**) Gluing bifurcation, $\mu \approx 1.2455$. (**c**) $\mu = 1.21$, one periodic orbit

> **Key Points from Chap. 6**
>
> - *Bifurcation theory* is concerned with the behaviour of differential equations including a parameter.
> - A *bifurcation* occurs if there is a change in the behaviour of the system as the parameter passes through a specific value.
> - A first-order differential equation $\dot{x} = f(x, \mu)$ has a *bifurcation* at $x = x_0$, $\mu = \mu_0$ if $f(x_0, \mu_0) = 0$ and $f_x(x_0, \mu_0) = 0$, with f_x changing sign.
> - A *bifurcation diagram* is a plot of $x(\mu)$, showing the existence and stability of fixed points as a function of the parameter μ.
> - For $\dot{x} = f(x, \mu)$ there are three standard types of bifurcation. Each of these has a *normal form*, which gives the behaviour of the equation near the bifurcation point.
> - At a *saddle–node bifurcation*, a stable fixed point collides with an unstable one and both disappear. The normal form is $\dot{x} = \mu - x^2$.
> - At a *transcritical bifurcation*, two fixed points cross each other and their stabilities switch over. The normal form is $\dot{x} = \mu x - x^2$.
> - At a *pitchfork bifurcation*, a symmetrical state becomes unstable and two symmetry-related fixed points are created. The normal form is $\dot{x} = \mu \pm x^3$.
> - These three main types of bifurcation also occur frequently in systems of higher order. This can be understood using the concept of a *centre manifold*, which reduces the system to first order near the bifurcation.
> - These three bifurcations are *stationary bifurcations*, where an eigenvalue passes through zero. For a system of order two or higher, a *Hopf bifurcation* can occur, where a fixed point has imaginary eigenvalues and a periodic orbit is created.
> - The bifurcations above are all *local*: they involve a change of stability of a fixed point and only influence a local region near that fixed point.
> - A *global bifurcation* does not require a fixed point to change its stability, and influences a large region of phase space. Typically, global bifurcations involve a change in the existence or structure of a periodic orbit.

Exercises

6.1 Find all the bifurcation points of $\dot{x} = 1 - \mu x + x^2$ and determine their type.

6.2 Starting from Eq. (6.6), rescale $x = a\hat{x}$, $\mu = c\hat{\mu}$, and show how a and c can be chosen to give the normal form (6.7). Under what conditions can this rescaling be done?

6.3 Show that the equation $\dot{x} = \mu^2 - x^2$ can be transformed into one of the three standard normal forms of Sect. 6.1 using the change of dependent variable from x to $y = x - \mu$. Hence state the type of bifurcation that occurs.

6.4 (a) Show that one of the fixed points of (6.9) is stable and the other is unstable.
(b) By changing variable from x to $y = x - a_1\mu$, show that (6.9) can be transformed into the normal form of a transcritical bifurcation (6.7).

6.5 Consider the equation $\dot{x} = (x^2 - 1)(\mu - x - x^2)$.

(a) Find the bifurcation points and identify the type of each bifurcation.
(b) Sketch the bifurcation diagram.
(c) Describe the behaviour of solutions as $t \to \infty$ with the initial condition $x = -1/2$ for (i) $\mu = -1$, (ii) $\mu = -3/16$, (iii) $\mu = 2$.

6.6 Find all the bifurcations that occur in $\dot{x} = \mu x + x^3 - x^5$, identify their type, and sketch the bifurcation diagram. Describe the behaviour as μ is gradually increased from -1 to $+1$ and then decreased from $+1$ back to -1.

6.7 Suppose that the saddle–node bifurcation (6.5) is modified to include a linear term βx. Show that by a suitable re-definition of x and μ this modified equation can be written in the form (6.5) (in other words, the modification does not qualitatively change the bifurcation).

6.8 Investigate the bifurcations that occur if the symmetry of a supercritical pitchfork bifurcation is broken by the addition of a quadratic term, $\dot{x} = \mu x + \delta x^2 - x^3$.

6.9 Consider the system

$$\dot{x} = y, \qquad \dot{y} = -y - x^2.$$

(a) Show that the fixed point at the origin has a zero eigenvalue and find the corresponding eigenvector.
(b) Writing the centre manifold as $y = a + bx + cx^2 + dx^3 + O(x^4)$, explain why $a = b = 0$.
(c) Using the fact that the centre manifold is invariant, find the coefficients c and d.
(d) Discuss the behaviour of solutions near the origin. Is the origin stable or unstable?

6.10 (a) Show that the origin is a neutrally stable fixed point of

$$\dot{x} = x^2 y - 2xy^2, \qquad \dot{y} = x - y + 2x^2 y,$$

and find the centre manifold up to the cubic terms.
(b) Hence show that the origin is stable.
(c) What property does this system have that forces certain terms in the Taylor expansion of the centre manifold to be zero?

6.11 Consider the problem of Exercise 6.10 with an extra term μx in the $\dot{x}$ equation.

(a) Show that the fixed point at the origin has a bifurcation when $\mu = 0$.

(b) For small μ, seek an extended centre manifold of the form

$$y = x + a\mu + b\mu x + c\mu^2 x + d\mu x^2 + ex^3 + O(4).$$

Explain why a and d must both be zero, and give the value of e based on your solution to Exercise 6.10.

(c) Use the fact that the ECM is an invariant curve to find b and c.
(d) Hence find the type of bifurcation that occurs at $\mu = 0$.

6.12 By considering the existence of fixed points, find the bifurcations that occur in

$$\dot{x} = \mu x - y + xy, \qquad \dot{y} = y - x^2.$$

6.13 Investigate the stability of all the fixed points of the system of Example 6.7, and find the value of μ at which there is a Hopf bifurcation.

6.14 Following the method of Example 6.8, find the relationship between the small parameter μ and the amplitude of the periodic orbit in the equation

$$\ddot{x} + (\dot{x}^2 - \mu)\dot{x} + x = 0,$$

and hence deduce whether the Hopf bifurcation is supercritical or subcritical.

6.15 Show that there is a Hopf bifurcation at $\mu = 0$ in the system

$$\dot{x} = \mu x - y + x(x^2 + y^2) + y(x^2 + y^2) - x(x^2 + y^2)^2,$$
$$\dot{y} = x + \mu y + y(x^2 + y^2) - x(x^2 + y^2) - y(x^2 + y^2)^2.$$

Use polar coordinates to show that the bifurcation is subcritical. Find all the periodic orbits that can exist, and hence deduce that there is a saddle–node bifurcation of periodic orbits at $\mu = -1/4$. Find the direction of the periodic orbits and explain what happens for the special case $\mu = 0$.

6.16 Follow the method of Example 6.9 to show that the gluing bifurcation in (6.32) occurs at approximately $\mu = 4\varepsilon/5$ for ε, μ small:

(a) After setting $\dot{x} = y$, find and classify the fixed points, and find the value of μ for which a Hopf bifurcation takes place.
(b) Show that the system is Hamiltonian in the limit $\mu \to 0$, $\varepsilon \to 0$ and find the Hamiltonian function, choosing the arbitrary constant so that $H(0, 0) = 0$.
(c) Find a homoclinic orbit in the Hamiltonian system and find where this curve crosses the x axis.
(d) Apply the Melnikov method and evaluate the integral. Hint: use the substitution $u = 1 - x^2/2$ to evaluate the integral.

Chapter 7
Difference Equations

So far, this book has been concerned with differential equations, where the independent and dependent variables are continuous. Another way of modelling systems that evolve in time is to use a discrete system, in which the state of the system is only recorded at certain times. This type of model might be used, for example, to describe how the summer population of a particular butterfly species varies from year to year, or how the maximum temperature changes from day to day.

These discrete dynamical systems can be written as a *difference equation*,

$$x_{n+1} - x_n = f(x_n). \tag{7.1}$$

Difference equations have many things in common with differential equations. For example, $x = x_*$ is a *fixed point* of (7.1) if $f(x_*) = 0$, so that $x_{n+1} = x_n$. This is analogous to the definition in Sect. 3.2 for a fixed point of a first-order differential equation. Also, for all the x_n to be defined, an initial condition, x_0, is needed, as for a differential equation.

Equation (7.1) gives a relationship between x_{n+1} and x_n, which defines a *first-order* difference equation. Difference equations can also occur at higher order. For example, the equation for the Fibonacci sequence,

$$x_{n+2} = x_{n+1} + x_n, \tag{7.2}$$

is a second-order difference equation. As for differential equations, a second-order difference equation can be written as two coupled first-order equations, so

$$x_{n+1} - x_n = f(x_n, y_n), \qquad y_{n+1} - y_n = g(x_n, y_n)$$

© The Author(s), under exclusive license to Springer Nature Switzerland AG 2025

P. C. Matthews, *Differential Equations, Bifurcations and Chaos*,

Springer Undergraduate Mathematics Series,

https://doi.org/10.1007/978-3-031-99543-9_7

is a second-order system, the discrete version of (5.1) and (5.2). This book will be mostly concerned with first-order difference equations, mainly because (7.1) can have far more complicated behaviour than a first-order differential equation.

One reason for studying difference equations is that the phenomenon of chaos, discussed in Chap. 8, is difficult to analyse in differential equations, but much easier to study in difference equations. Another reason is that many mathematical problems that can't be solved exactly can be solved approximately by an iterative process using a difference equation. A good example of such an iterative method is Newton's method, described in Sect. 7.1.

Difference equations arise directly from differential equations in two different contexts. The first is when a differential equation is discretised in order to find an approximate numerical solution, see Sect. 7.2. The second occurs when studying the stability of a periodic orbit, as described in Sect. 7.3.

Remark 7.1 In most textbooks, discrete dynamical systems are written as

$$x_{n+1} = F(x_n), \tag{7.3}$$

rather than (7.1). Of course, the two forms are equivalent, since $F(x_n) = f(x_n) + x_n$. There are some advantages of using the form (7.1) in a book where the main focus is on differential equations. For example, as mentioned above, the definition of a fixed point is the same. Similarly, the three main types of bifurcation occur in the same way with the same normal form. Also, the two important numerical methods described in the following sections are written in the form (7.1). However, there are also some advantages of using the form (7.3), as we shall see in the final chapter.

7.1 Newton's Method

An important and useful example of a difference equation is Newton's method for the solution of $G(x) = 0$. Starting from an initial guess x_0, Newton's method is

$$x_{n+1} = x_n - \frac{G(x_n)}{G'(x_n)}. \tag{7.4}$$

Suppose that x_* is a solution, so $G(x_*) = 0$. Then x_* is a fixed point of (7.4). If x_0 is close to x_*, then the method converges rapidly to x_*, but if the initial guess is poor, the method may take a long time to converge, or may not even converge at all (see Example 7.10 and Exercise 7.11).

Example 7.1 Use Newton's method to estimate $\sqrt{10}$.

Newton's method solves $G(x) = 0$, so in this case $G(x) = x^2 - 10$. The method (7.4) is then

$$x_{n+1} = x_n - \frac{x_n^2 - 10}{2x_n}.$$

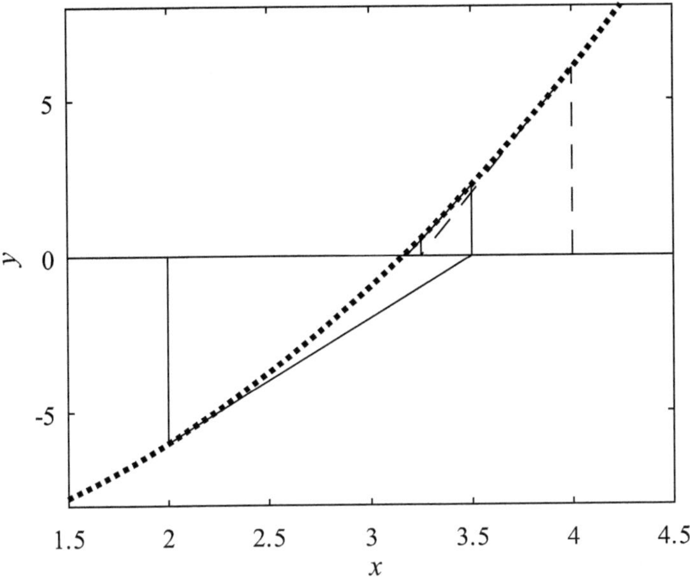

Fig. 7.1 Newton's method (7.4) applied to $G(x) = x^2 - 10 = 0$. The dotted curve is $G(x)$. The dashed lines show Newton's method with starting value $x_0 = 4$, and the solid lines show the method for $x_0 = 2$. In both cases, the method converges quickly to the solution $\sqrt{10}$

Since $\sqrt{9} = 3$, $x_0 = 3$ is a good choice of starting value. Then $x_1 = 3 + 1/6 = 19/6$ and $x_2 = 19/6 - (1/36)/(19/3) = 3.16228\ldots$, which is accurate to five decimal places after just two steps. The remarkably fast convergence of Newton's method is investigated in Example 7.3.

Figure 7.1 shows how the method works geometrically, for two different starting values. Starting from the initial guess x_0, the method finds the value of the function $G(x_0)$, shown by a vertical line from the x axis to the curve $G(x)$. Then the tangent line to the curve at this point is followed until it crosses the x axis, which gives the next step of the iteration, x_1. This process is repeated to generate a sequence x_n, which usually converges quickly to a root of $G(x)$.

7.2   Numerical Methods for Differential Equations

Methods for investigating solutions of differential equations include analytical methods, as described in Chap. 2, qualitative methods, which are the main focus of this book, and numerical methods, briefly described in this section.

Analytical methods work by first finding a general solution, including a constant, and then applying the initial condition to find the value of the constant. Numerical methods work the other way round, starting from the initial condition. Suppose that the differential equation and initial condition are

$$\dot{x} = f(x), \qquad x(0) = a. \tag{7.5}$$

Substituting the initial condition into the differential equation gives $\dot{x}(0) = f(a)$, which gives the initial slope of the function $x(t)$, in other words, the tangent to the curve at this point. From this we can take a small step along the tangent line, and then repeat this process, building up a sequence of points that form an approximate solution to the differential equation.

One way to derive the simplest type of numerical method is to start from the definition of the derivative,

$$\frac{dx}{dt} = \lim_{h \to 0} \frac{x(t+h) - x(t)}{h}, \tag{7.6}$$

provided that this limit exists. Now suppose that t and x are discretised by choosing a fixed value of h, setting $t_n = nh$, for $n = 0, 1, 2, 3 \ldots$, and defining $x_n = x(t_n)$. Then if h is small, an approximate discrete form of (7.5) is

$$\frac{x_{n+1} - x_n}{h} \approx f(x_n), \qquad x_0 = a,$$

which is a difference equation. Now let X_n be a numerical approximation of x_n. Then the formula

$$X_{n+1} = X_n + hf(X_n) \tag{7.7}$$

is known as *Euler's method*.[1] This formula can be thought of as the first two terms in a Taylor series for $x(t)$, which indicates that the error in taking one step of the method is proportional to h^2. For solving a differential equation over a finite interval, say $0 < t < 1$, the number of steps needed is $1/h$, so the total error, assuming that the errors add up rather than cancelling out, is the error per step multiplied by the number of steps, which is of order h.

Example 7.2 Apply Euler's method to the logistic equation $\dot{x} = x - x^2$, with $x(0) = 2$ and $h = 0.2$.

Euler's method with $h = 0.2$ is $X_{n+1} = X_n + 0.2(X_n - X_n^2)$. Starting from $X_0 = 2$ gives $X_1 = 2 + 0.2(2 - 4) = 1.6$, then $X_2 = 1.6 + 0.2(1.6 - 1.6^2) = 1.408$.

---

[1] This is one of many things named after the eighteenth century Swiss mathematician Leonhard Euler (pronounced "Oiler").

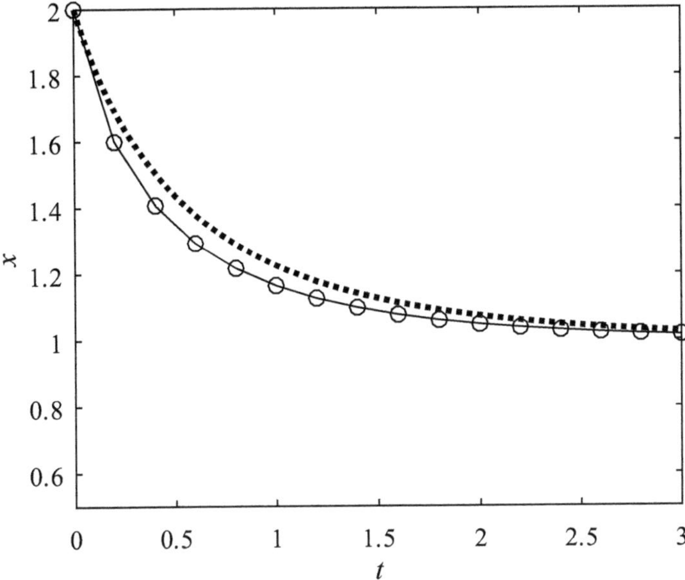

Fig. 7.2 Euler's method (7.7) applied to $\dot{x} = x - x^2$, with $x(0) = 2$ and $h = 0.2$. The dotted curve is the exact solution. The straight line segments joined by circles show the steps of the numerical method

To four decimal places, $X_3 = 1.2931$ and $X_4 = 1.2173$, and continuing further, $X_n \to 1$ as $n \to \infty$, which is the correct stable fixed point of the differential equation. The numerical approximation and the true solution are shown in Fig. 7.2.

Remark 7.2 There are many more complicated numerical methods that are more accurate than Euler's method. However, Euler's method has some useful features. It is clear from (7.7) that there is an exact one-to-one correspondence between fixed points of the differential equation and fixed points of the difference equation: $f(X_n) = 0$ if and only if $X_{n+1} = X_n$. This nice property does not hold for some of the more sophisticated methods (see Exercise 7.13). Although there is an error per step of order h^2, the error is zero at a fixed point, so if we only want to use a numerical method to find a stable fixed point of a differential equation, Euler's method is as good as any other, see Fig. 7.2. Finally, Euler's method is very simple to code, see the pseudocode below (the term *pseudocode* means a computer code that is not written in any specific programming language, but is easy to read and understand and implement in a computer language of the reader's choice).

Pseudocode for Euler's Method

```
# Euler's method for dx/dt = x - x^2
h = 0.2 # step size
t = 0 # start time
nsteps = 50 # number of steps used
x = 2 # initial condition
for n from 1 to nsteps
    t = t + h
    x = x + h*(x - x^2)
    print t, x
end for
```

7.3 Stability of Periodic Orbits

The second example where difference equations arise directly from differential equations is in the investigation of the stability of periodic orbits. Stable and unstable periodic orbits were discussed in Sect. 5.8, in terms of whether nearby solutions approach the periodic orbit or move away from it. But what are the possible types of instability of periodic orbits and what are the sorts of bifurcations that occur at these instabilities?

To answer these questions, a useful concept is the *Poincaré section*. In the simplest case, suppose that a second-order system has a periodic orbit, and consider a fixed straight line that intersects the periodic orbit, as shown in Fig. 7.3, where the line has been chosen to be a line of constant y, $y = y_0 = -1$. Now consider a trajectory that starts on this line, near the periodic orbit, where x has the value x_0. Suppose that, as in the figure, $\dot{y} > 0$ at (x_0, y_0). This trajectory loops around the periodic orbit, crosses the line $y = y_0$ at a point where $\dot{y} < 0$, and then crosses the line $y = y_0$ again with $\dot{y} > 0$. The value of x where this occurs can be labelled x_1. Repeatedly following the points on the trajectory where $y = y_0$ with $\dot{y} > 0$ gives an infinite sequence of discrete x values, $x_0, x_1, x_2, x_3\ldots$, and since these values are related to each other by the underlying differential equations, the change in the value of x between each crossing depends on the previous value of x, so we have a difference equation of the form (7.1). In the case shown in Fig. 7.3, the sequence of points x_n is decreasing and approaches a fixed point corresponding to the periodic orbit as $n \to \infty$, and the orbit is stable. If the orbit was unstable, the sequence of points x_n would move away from the periodic orbit.

Alternatively, we could drop the condition that crossings of the line $y = y_0$ must occur with $\dot{y} > 0$. Then, in the example shown in Fig. 7.3, the sequence x_n includes the crossing points where $x < 0$. In that case, as $n \to \infty$, the sequence x_n approaches a state where x_n alternates between two values, corresponding to the two points where the stable periodic orbit crosses $y = y_0$.

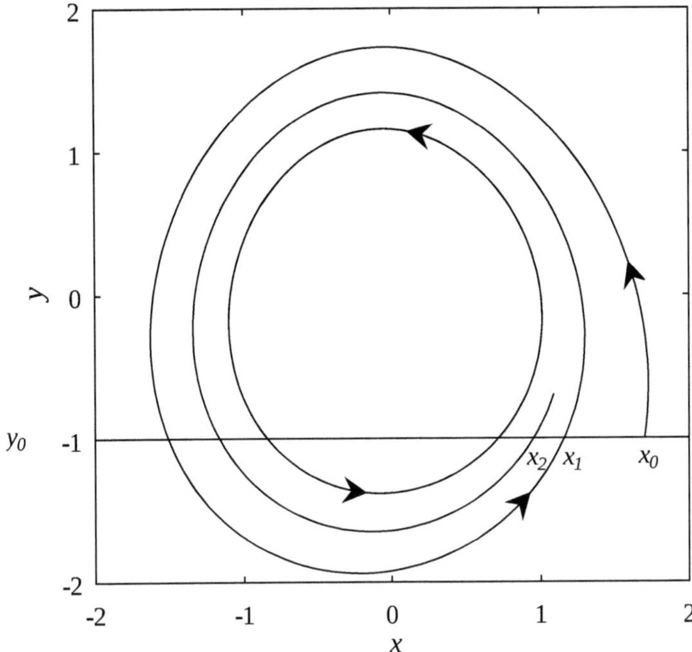

Fig. 7.3 A trajectory near a periodic orbit crosses the line $y = y_0$ with $\dot{y} > 0$ repeatedly. This generates a sequence of points $x_0, x_1, x_2 \ldots$

The purpose of this is not to construct the difference equation, because to do this we would have to be able to solve the differential equation. But this theoretical construction shows that instabilities of periodic orbits are very closely related to instabilities of difference equations. The fixed line where the crossings are measured is known as a *Poincaré section*, and the corresponding difference equation is sometimes called a *Poincaré map*.

The same approach can be used to study the stability of periodic orbits in higher-order systems of differential equations. For example, for a third-order system, the Poincaré section is a two-dimensional plane, and the difference equation is second order.

7.4   Fixed Points and Periodic Points in Difference Equations

Definition 7.1 The point x_* is a *fixed point* of the difference equation (7.1) if and only if $f(x_*) = 0$.

Definition 7.2 The sequence of points $\{x_1, x_2, \ldots x_p\}$, for $p > 1$, form a *periodic solution* of (7.1) with period p, or a *period-p* solution, if and only if

$$x_{j+1} - x_j = f(x_j) \quad \text{for} \quad 1 \leq j \leq p - 1 \quad \text{and} \quad x_1 - x_p = f(x_p),$$

provided that all the points x_j, $j = 1 \ldots p$ are distinct. Any point contained in a periodic solution is a *periodic point*.

So a point x_1 is periodic with period p, or p-*periodic*, if $x_{p+1} = x_1$ and $x_j \neq x_1$ for $1 < j < p + 1$.

The condition that the points in a periodic solution must be distinct is needed to make the definition of the period p precise. Consider the sequence $x_n = (-1)^n$. Clearly this sequence has period two. It is also true that $x_{k+4} = x_k$ and $x_{k+6} = x_k$, but according to Definition 7.2, x_n does not have period four or six. This means that if we find a solution of (7.1) with the property that $x_{k+4} = x_k$, we need to check carefully whether the solution has period four, since it could have period two, or could be a fixed point.

7.5 Stability and Instability in Difference Equations

The stability of a fixed point of a difference equation can be investigated in the same way as for differential equations, see Sect. 3.2.1.

Suppose that x_* is a fixed point of (7.1), so $f(x_*) = 0$. Now let $x_n = x_* + y_n$, and assume that y_n is small, so that x_n is close to the fixed point. Then

$$y_{n+1} - y_n = x_{n+1} - x_n = f(x_* + y_n) = f(x_*) + f'(x_*)y_n + O(y_n^2), \tag{7.8}$$

and since $f(x_*) = 0$, the linearised problem for y_n is the difference equation

$$y_{n+1} - y_n = f'(x_*)y_n. \tag{7.9}$$

The solution of this linear difference equation is

$$y_n = (1 + f'(x_*))^n y_0, \tag{7.10}$$

which is the discrete version of the exponential solution (3.7). The behaviour of (7.10) depends on the value of $f'(x_*)$.

- If $f'(x_*) > 0$ then y_n increases in magnitude, so the fixed point is unstable, as for a differential equation.
- If $-2 < f'(x_*) < 0$ then at each step y_n is multiplied by a factor between 1 and -1, so $y_n \to 0$ as $n \to \infty$ and the fixed point is stable.
- If $f'(x_*) < -2$, the multiplying factor in (7.10) is less than -1, so y_n oscillates in sign and increases in magnitude, in which case the fixed point is unstable.

- In the special case $f'(x_*) = -1$, the linearised equation (7.9) gives $y_{n+1} = 0$ and the fixed point is said to be *superstable*. The convergence to the fixed point is faster in this case. The behaviour of y_n depends on the next term in the Taylor series (7.8), and as long as $f''(x_*) \neq 0$, $y_{n+1}/y_n^2 \to f''(x_*)/2 \neq 0$ as $n \to \infty$, which is known as *quadratic convergence*. Since y_{n+1} is proportional to y_n^2, the sequence y_n converges to 0 very rapidly.
- The borderline cases $f'(x_*) = 0$ and $f'(x_*) = -2$ correspond to bifurcations. These cases are studied in Sects. 7.6 and 7.7 respectively.

In summary, a fixed point x_* of (7.1) is stable if

$$-2 < f'(x_*) < 0, \tag{7.11}$$

with instabilities occurring if either $f'(x_*) > 0$ or $f'(x_*) < -2$. The fixed point is *superstable* if

$$f'(x_*) = -1. \tag{7.12}$$

Example 7.3 Show that a fixed point of Newton's method is superstable.

Newton's method (7.4) is a difference equation of the form (7.1) with $f(x) = -G(x_n)/G'(x_n)$. A point x_* is a fixed point of the method if and only if $G(x_*) = 0$. The stability of a fixed point x_* depends on

$$f'(x_*) = \frac{G(x_*)G''(x_*) - G'(x_*)G'(x_*)}{G'(x_*)^2} = -1,$$

so x_* is a superstable fixed point. This means that Newton's method converges rapidly, at least quadratically, as observed in Example 7.1. If $G''(x_*) = 0$ then $f''(x_*) = 0$ and convergence is at least cubic.

The bifurcations corresponding to each of the stability boundaries in (7.11) are investigated in the following two sections. Since the first type is very similar to bifurcations in first-order differential equations, this case is only covered briefly in Sect. 7.6. More time is given to the second case (Sect. 7.7), which is a new type of instability that does not occur for the fixed points of differential equations.

7.6 Stationary Bifurcations of Difference Equations

Consider a difference equation including a parameter μ,

$$x_{n+1} - x_n = f(x_n, \mu). \tag{7.13}$$

As the parameter μ varies, the stability of a fixed point may change. We will use the term *stationary bifurcation*, as for differential equations, to describe a point (x_*, μ_*) where a fixed point crosses a stability boundary with $\frac{\partial f}{\partial x}(x_*, \mu_*) = 0$.

Stationary bifurcations for difference equations can be investigated using a Taylor series expansion exactly as for differential equations. Since the calculations are virtually identical to those of Sect. 6.1, only the results are given here. The three main types of bifurcation are the same, with the same normal forms except that x is replaced by x_n and $\dot{x}$ becomes $x_{n+1} - x_n$. As in the differential equation case, a shift of coordinates is carried out when writing down the normal forms, so that the bifurcation point is at $x_* = 0$, $\mu_* = 0$.

If there are no constraints on the partial derivatives of f, the bifurcation is in general a saddle–node or turning-point bifurcation. Near the bifurcation point, where x_n and μ are small, the difference equation can be reduced to the normal form

$$x_{n+1} - x_n = \mu - x_n^2. \tag{7.14}$$

There are no fixed points for $\mu < 0$, and two fixed points for $\mu > 0$, a stable one at $x* = \sqrt{\mu}$ and an unstable one at $x_* = -\sqrt{\mu}$. To make the phrase "in general" more precise, (7.14) applies if f_μ and f_{xx} are both non-zero at the bifurcation point. Note that this is all the same as for a saddle–node bifurcation in a first-order differential equation, see Sect. 6.1.1.

If it is known that a fixed point exists for all values of μ, then in general the bifurcation is transcritical, with two fixed points crossing over and switching stabilities, and the normal form is

$$x_{n+1} - x_n = \mu x_n - x_n^2. \tag{7.15}$$

The two fixed points of (7.15) are $x_* = 0$, which is stable for $\mu < 0$ but unstable for $\mu > 0$, and $x_* = \mu$, which is stable for $\mu > 0$ and unstable for $\mu < 0$.

If the difference equation has the symmetry $x_n \leftrightarrow -x_n$, then $f(x_n, \mu)$ must be an odd function of x_n, implying that there is a fixed point at $x_* = 0$ for all μ. A stationary bifurcation from this fixed point is a pitchfork, at which the stability of the $x_* = 0$ fixed point changes and two new fixed points are created. There are two possibilities for the normal form,

$$x_{n+1} - x_n = \mu x_n \pm x_n^3. \tag{7.16}$$

In the $-$ case, known as supercritical, the two extra fixed points exist for $\mu > 0$ and are stable. For the $+$ case, called subcritical, the two fixed points are unstable and exist for $\mu < 0$.

For each of these bifurcations there is a corresponding bifurcation diagram, which is a plot of the stable and unstable fixed points x_* as a function of μ, identical to those for differential equations shown in Sect. 6.1. The stability conditions given above are only valid within the range of validity of the normal form, that is, when x_n and μ are both small.

Example 7.4 Find and classify the bifurcations in the difference equation (7.13) for $f(x, \mu) = \cos x + \mu$.

At a fixed point x_*, μ and x_* are related by $\mu = -\cos x_*$. The stability depends on $f_x = -\sin x$. Since $-1 \leq \sin x \leq 1$, the second type of instability mentioned in the previous section ($f_x < -2$) is not possible, so only stationary bifurcations with $f_x = 0$ can occur. These bifurcations happen when $\sin x = 0$, so $x = n\pi$ and $\mu = -\cos x = (-1)^{n+1}$. Consider the bifurcation at $x = 0$, $\mu = -1$. Near this point, x is small and $\mu = -1 + \delta$, where δ is small. Expanding the difference equation near the bifurcation point,

$$x_{n+1} - x_n = 1 - \frac{x_n^2}{2} + O(x_n^4) + (-1 + \delta) = \delta - \frac{x_n^2}{2} + O(x_n^4).$$

Hence the bifurcation is a saddle–node, and it can be written in the normal form (7.14) after a slight rescaling. Alternatively, checking that $f_\mu = 1$ and $f_{xx} = -\cos x$ are both non-zero at the bifurcation point shows that the bifurcation is a saddle–node. Similarly, it can easily be shown that all the other bifurcations are saddle–nodes.

Since the stability of fixed points of difference equations is closely related to the stability of periodic orbits in differential equations, as shown in Sect. 7.3, periodic orbits can have each of these three types of bifurcation. At a saddle–node bifurcation, a stable periodic orbit merges with an unstable one and both disappear (see Exercise 6.15 for an example). A transcritical bifurcation occurs when two periodic orbits cross through each other. A pitchfork bifurcation can occur when a symmetrical periodic orbit becomes unstable and two asymmetric periodic orbits are created.

7.7   Period-Doubling Bifurcations in Difference Equations

Consider the difference equation with a parameter (7.13) near the point of instability where $\frac{\partial f}{\partial x}(x_*, \mu_*) = -2$. At this point, (7.10) shows that according to the linearised difference equation, the perturbation y_n obeys $y_n = (-1)^n y_0$, so solutions alternate in sign, flipping back and forth between y_0 and $-y_0$. This is called a *period-doubling* or *flip* bifurcation. By including nonlinear terms and allowing μ to vary slightly from the bifurcation point $\mu = \mu_*$, the behaviour near this bifurcation point can be understood.

This bifurcation can be studied by using a Taylor series method similar to that used in Sect. 6.1, but this calculation is quite cumbersome, so only the result is given here. The normal form of the period-doubling bifurcation can be written as

$$x_{n+1} - x_n = -2x_n - \mu x_n \pm x_n^3. \tag{7.17}$$

This normal form is similar to the pitchfork bifurcation (7.16), as only linear and cubic terms appear and there are two different forms, depending on the sign of the cubic term. There is a fixed point at $x_* = 0$ that exists for all μ. In applications, this fixed point generally lies on a curve in the (x, μ) plane, but this has been moved to $x_* = 0$ by coordinate transformations to get (7.17). This fixed point is stable if $\mu < 0$ and unstable if $\mu > 0$, with a period-doubling bifurcation at $\mu = 0$. If there is another fixed point x_*, then $0 = -2 - \mu \pm x_*^2$. Although this may have solutions, they are not relevant, because as with all normal form equations, (7.17) only applies near the bifurcation point, where μ and x_n are both small, and there are no solutions that satisfy this condition.

However, (7.17) does have a solution in which μ and x_n are small and $x_{n+1} = -x_n \neq 0$. This is called the *period-two* solution. If this condition holds then, from (7.17),

$$\mu = \pm x_n^2, \tag{7.18}$$

so this solution exists for $\mu > 0$ in the $+$ case and $\mu < 0$ in the $-$ case. The stability of the period-two state can be found by setting x_n to the solution of (7.18) plus a small perturbation δ_n in (7.17) and then linearising. Then

$$\delta_{n+1} - \delta_n = -2\delta_n - \mu\delta_n \pm 3x_n^2\delta_n,$$

and substituting for x_n^2 using (7.18) gives

$$\delta_{n+1} = (-1 + 2\mu)\delta_n.$$

If $\mu > 0$ then $\delta_n \to 0$ as $n \to \infty$, showing that the period-two state is stable, but if $\mu < 0$ it is unstable (remember that μ is assumed to be small).

In summary, in the $+$ case, the period-two state exists for $\mu > 0$ and is stable, but in the $-$ case it exists for $\mu < 0$ and is unstable.

The bifurcation diagrams for the two cases look the same as for the pitchfork bifurcation, see Fig. 6.4. The difference is that the non-zero branches are period-two states where the solution switches to and fro between the two branches. If the period-two solution is stable, it exists where the fixed point is unstable and the period-doubling bifurcation is supercritical. If it is unstable, it exists when the fixed point is stable and the bifurcation is subcritical.

In examples, there is no point in doing the complicated algebra to convert the bifurcation to the normal form (7.17). It is sufficient to find the point where $f_x = -2$ and then, if possible, find the period-two solution explicitly. This can be done by labelling the two states of the period-two solution as x_1 and x_2, with $x_2 - x_1 =$

$f(x_1)$, $x_1 - x_2 = f(x_2)$, $x_1 \neq x_2$, and then solving for x_1 and x_2, as in the following example.

Example 7.5 Find the location of the period-doubling bifurcation in

$$x_{n+1} - x_n = \mu - x_n^2. \tag{7.19}$$

Find the period-two solution explicitly, and hence show that the period-doubling bifurcation is supercritical.

The difference equation (7.19) is of the form (7.13) with $f(x, \mu) = \mu - x^2$. Fixed points x_* exist if $\mu > 0$, with $\mu = x_*^2$. In fact, (7.19) is the normal form for the saddle–node bifurcation (7.14), but this example is concerned with the behaviour of (7.19) away from the bifurcation at $\mu = 0$. For the fixed point at $x = -\sqrt{\mu}$, $f_x = -2x = 2\sqrt{\mu} > 0$ so this point is always unstable. For $x = \sqrt{\mu}$, $f_x = -2\sqrt{\mu} < 0$. Recall from (7.11) that the stability condition is $-2 < f_x(x_*) < 0$, so this point is stable if $\mu < 1$ and unstable if $\mu > 1$, with a period-doubling bifurcation at $\mu = 1$.

To find the period-two solution, label its two points as x_1 and $x_2 \neq x_1$. We then need to solve the nonlinear simultaneous equations

$$x_2 - x_1 = \mu - x_1^2,$$

$$x_1 - x_2 = \mu - x_2^2.$$

Subtracting these equations gives

$$2(x_2 - x_1) = x_2^2 - x_1^2 = (x_2 + x_1)(x_2 - x_1).$$

One solution is $x_1 = x_2$, but this is simply the fixed point of (7.19), not the period-two solution we are looking for. Cancelling this factor gives $x_1 + x_2 = 2$, and using this to eliminate x_2 from the first of the simultaneous equations shows that x_1 obeys the quadratic

$$x_1^2 - 2x_1 + 2 - \mu = 0 \quad \Rightarrow \quad (x_1 - 1)^2 + 1 - \mu = 0 \quad \Rightarrow \quad x_1 = 1 \pm \sqrt{\mu - 1}.$$

Hence the period-two solution exists for $\mu > 1$, which is where the fixed point is unstable. Therefore the period-doubling bifurcation is supercritical and the period-two solution is stable near the bifurcation.

The next example uses a different method to find the period-two solution, and also calculates its stability.

Example 7.6 For the difference equation (7.19), find a difference equation for $x_{n+2} - x_n$. Hence find the period-two solution as a fixed point of this equation, investigate its stability, and sketch the bifurcation diagram.

From (7.19), $x_{n+2} - x_{n+1} = \mu - x_{n+1}^2$. Adding this equation to (7.19) gives

$$x_{n+2} - x_n = 2\mu - x_n^2 - x_{n+1}^2 = 2\mu - x_n^2 - (x_n + \mu - x_n^2)^2. \tag{7.20}$$

This is now in the form required, a second-order difference equation for $x_{n+2} - x_n$. The reason for doing this is that a period-two solution of (7.19) has $x_{n+2} = x_n$, so it is a fixed point of (7.20), obeying

$$2\mu - x_n^2 - (x_n + \mu - x_n^2)^2 = 0. \tag{7.21}$$

The fixed points of the original difference equation (7.19), where $x_*^2 = \mu$, will also appear as fixed points of (7.20), so these need to be factored out. Knowing in advance that $(\mu - x_n^2)$ must be a factor of (7.21) helps to simplify the algebra. Now we expand (7.21), but keep factors of $(\mu - x_n^2)$ intact:

$$2\mu - x_n^2 - x_n^2 - 2x_n(\mu - x_n^2) - (\mu - x_n^2)^2 = 0.$$

The expected factor $(\mu - x_n^2)$ can now be taken out, leaving

$$2 - 2x_n - \mu + x_n^2 = 0,$$

which is the quadratic found in Example 7.5, so the result is the same.

This method also gives the stability of the period-two solution, as it is a fixed point of (7.20). We have shown that (7.20) can be factorised as

$$x_{n+2} - x_n = (\mu - x_n^2)(2 - 2x_n - \mu + x_n^2).$$

On the period-two solution, where the second bracket is zero, the x-derivative of the right-hand side is

$$(\mu - x_n^2)(-2 + 2x_n) = (2 - 2x_n)(-2 + 2x_n) = -4(x_n - 1)^2 = -4(\mu - 1),$$

which is negative because the period-two solution only exists for $\mu > 1$. The period-two state is stable if $-2 < -4(\mu - 1) < 0$. Near the period-doubling bifurcation, where μ is just greater than 1, this condition is satisfied, so the period-two solution is stable and hence the period-doubling bifurcation is supercritical. But at $\mu = 3/2$ it becomes unstable. Hence the fixed point of (7.20) has a period-doubling bifurcation at $\mu = 3/2$, so at this point a period-four solution of (7.19) is created. What happens for larger μ will be investigated in Chap. 8. The bifurcation diagram for (7.19) is shown in Fig. 7.4.

The method used in Example 7.6 can be applied to the general first-order difference equation (7.1). To find the second-order difference equation relating x_{n+2} and x_n, we add $x_{n+1} - x_n = f(x_n)$ and $x_{n+2} - x_{n+1} = f(x_{n+1})$ to get

$$x_{n+2} - x_n = f(x_n) + f(x_{n+1}) = f(x_n) + f(x_n + f(x_n)). \tag{7.22}$$

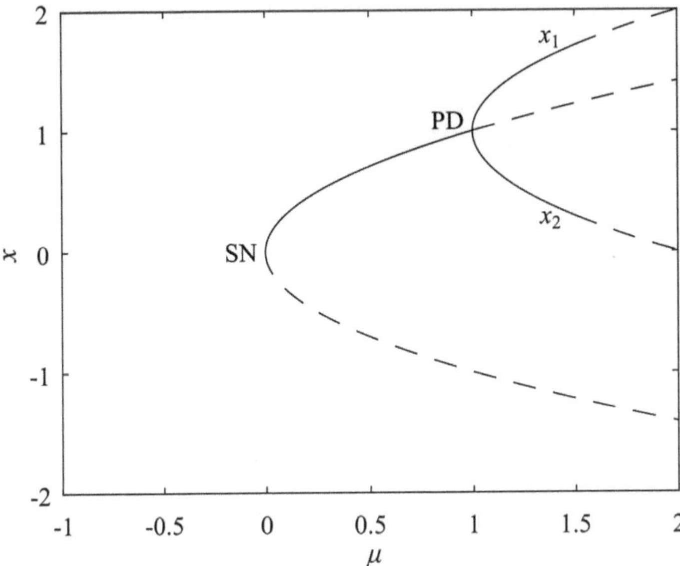

Fig. 7.4 Bifurcation diagram for (7.19). A stable fixed point and an unstable one are created at a saddle–node bifurcation (SN) at $\mu = 0$. The stable branch becomes unstable at a period-doubling bifurcation (PD) at $\mu = 1$, where a period-two solution labelled x_1, x_2 is created, which is stable for $1 < \mu < 3/2$

A period-two solution of (7.1) has $x_{n+2} = x_n$, so it is a fixed point of (7.22). So for a period-two solution with points x_1 and $x_2 \neq x_1$, we need $x_2 - x_1 = f(x_1)$ and $f(x_1) + f(x_2) = 0$. The stability of a period-two state is determined by the x-derivative of $f(x) + f(x + f(x))$, evaluated at $x = x_1$, which is, using the chain rule for the derivative of a function of a function,

$$f'(x_1) + f'(x_1 + f(x_1))(1 + f'(x_1)) = f'(x_1) + f'(x_2) + f'(x_1)f'(x_2).$$

The period-two state is stable if this quantity, which we will call $K(x_1, x_2)$, lies between -2 and 0, so the stability condition is

$$-2 < K(x_1, x_2) = f'(x_1) + f'(x_2) + f'(x_1)f'(x_2) < 0. \tag{7.23}$$

As we would expect, the stability function $K(x_1, x_2)$ is symmetrical, that is, $K(x_1, x_2) = K(x_2, x_1)$.

Example 7.7 Use (7.23) to investigate the stability of the period-two solutions of $x_{n+1} - x_n = 2/x_n - \mu$.

For $x_{n+1} - x_n = 2/x_n - \mu$, there is a fixed point at $x_* = 2/\mu$, which is stable if $-2 < -2/x_*^2 < 0$, so $-2 < -\mu^2/2 < 0$. The fixed point does not have a stationary bifurcation. It is stable for $-2 < \mu < 2$ and there are period-doubling bifurcations at $x_* = \pm 1$, $\mu = \pm 2$. For a period-two solution (x_1, x_2),

$$x_2 - x_1 = \frac{2}{x_1} - \mu \quad \text{and} \quad x_1 - x_2 = \frac{2}{x_2} - \mu. \tag{7.24}$$

Subtracting these two equations,

$$2(x_2 - x_1) = \frac{2}{x_1} - \frac{2}{x_2} = \frac{2(x_2 - x_1)}{x_1 x_2},$$

and factoring out the fixed point solution $x_2 = x_1$ shows that $x_1 x_2 = 1$ for a period-two solution. Eliminating $x_2 = 1/x_1$ from the first of (7.24) gives

$$x_1^2 - \mu x_1 + 1 = 0. \tag{7.25}$$

Solutions exist if $\mu^2 > 4$, which is consistent with period-doubling bifurcations occurring at $\mu = \pm 2$. The stability function in (7.23) is

$$K(x_1, x_2) = -\frac{2}{x_1^2} - \frac{2}{x_2^2} + \frac{4}{x_1^2 x_2^2} = \frac{4 - 2x_1^2 - 2x_2^2}{x_1^2 x_2^2} = \frac{4 - 2(x_1 + x_2)^2 + 4x_1 x_2}{x_1^2 x_2^2}.$$

By writing it in this form, we can evaluate $K(x_1, x_2)$ using the rules for the sum and product of the roots of the quadratic (7.25), without having to write down x_1 and x_2 explicitly.[2] From (7.25), $x_1 + x_2 = \mu$ and $x_1 x_2 = 1$, so $K(x_1, x_2) = 8 - 2\mu^2$. The period-two states are stable if $-2 < K(x_1, x_2) < 0$, so $4 < \mu^2 < 5$. There is a secondary period-doubling bifurcation, leading to a period-four solution, when $\mu = \pm\sqrt{5}$.

Example 7.8 Suppose that Euler's method is applied to a nonlinear differential equation with a stable fixed point at $x = x_*$. Show that x_* is also a stable fixed point of the numerical method if the step size h is small, and that as h is increased, the fixed point becomes unstable at a period-doubling bifurcation.

If the differential equation $\dot{x} = f(x)$ has a stable fixed point at $x = x_*$ then $f(x_*) = 0$ and $f'(x_*) < 0$. Euler's method with a step size $h > 0$ is

$$X_{n+1} - X_n = hf(X_n). \tag{7.26}$$

As noted in Remark 7.2, a fixed point of this difference equation is exactly a fixed point of the differential equation; there is no numerical error in the value of a fixed point. Using (7.11), the fixed point at $x = x_*$ is stable if $-2 < hf'(x_*) < 0$. Since $h > 0$ and $f'(x_*) < 0$, the right-hand inequality is always true, so a stationary bifurcation cannot occur. But there is a period-doubling bifurcation at $h = -2/f'(x_*)$, with stability for $h < -2/f'(x_*)$. So the method is stable for

---

[2] Matthews's law: Never use the quadratic formula unless you absolutely have to.

small h but becomes unstable at a period-doubling bifurcation if h is too large. See Exercise 7.8 for a specific example of this instability.

Remark 7.3 Example 7.8 gives another example of different areas of mathematics linking up. In numerical analysis, the behaviour of (7.26) as h gets too large is known as the *numerical instability* of Euler's method. In dynamical systems, the same thing is called a period-doubling bifurcation.

7.8   Symmetry Effects in Difference Equations

A recurring theme in this book is the way in which symmetry in the equations, usually arising from a symmetry of the underlying system being modelled, has a significant influence on the nonlinear dynamics of the system. Symmetry in a system of differential equations can lead to the presence of an invariant line (Sect. 5.6) and can force the existence of a fixed point. It can also constrain the form of a stationary bifurcation, forcing it to be a pitchfork (Sect. 6.1.3), and the type of a global bifurcation (Fig. 6.10).

Consider a first-order difference equation with a parameter (7.13) in the case where $f(x, \mu)$ is an odd function of x, so that the equation has a symmetry $x_n \leftrightarrow -x_n$. To be more precise, let $\hat{x}_n = -x_n$. Then the difference equation for $\hat{x}_n$ is

$$\hat{x}_{n+1} - \hat{x}_n = -x_{n+1} + x_n = -f(x_n, \mu) = f(-x_n, \mu) = f(\hat{x}_n, \mu),$$

so the equation for $\hat{x}_n$ is exactly the same as the equation for x_n. Hence, for any fixed point x_*, there is also a corresponding fixed point $-x_*$. For any period-two solution (x_1, x_2), there is also a period-two solution $(-x_1, -x_2)$, which may be the same solution (if $x_2 = -x_1$) or not (if $x_2 \neq -x_1$). Also, 0 must be a fixed point for all μ, because $f(0, \mu) = 0$.

The fixed point at $x = 0$ cannot have a saddle–node bifurcation or a transcritical bifurcation, since neither of these satisfy the sign-change symmetry. But it can undergo a pitchfork bifurcation, at a value of μ where the partial derivative $f_x(0, \mu)$ is zero, leading to two symmetry-related fixed points. It can also have a period-doubling bifurcation, if μ obeys $f_x(0, \mu) = -2$, creating a period-two cycle between two points x_1 and $x_2 = -x_1$.

There is one more interesting result that can be deduced in this case. Recall that in Examples 7.6 and 7.7 a period-two cycle had a secondary period-doubling bifurcation, creating a period-four cycle. But in the symmetrical case considered here, this cannot happen. A symmetric period-two cycle $(x_1, -x_1)$ cannot have a secondary period-doubling bifurcation. To show this, consider the stability condition (7.23) for the period-two cycle regarded as a fixed point of the difference equation for $x_{n+2} - x_n$, expressed slightly differently since f is now a function of two variables, x and μ:

$$-2 < f_x(x_1) + f_x(x_2) + f_x(x_1)f_x(x_2) < 0.$$

In the symmetrical case, $x_2 = -x_1$, and because f is an odd function of x, f_x is an even function, so $f_x(x_1) = f_x(x_2)$. Then the stability condition is

$$-2 < 2f_x(x_1) + f_x(x_1)^2 = f_x(x_1)(2 + f_x(x_1)) < 0. \tag{7.27}$$

Now for any real number r, $(1 + r)^2 \geq 0 \implies r(2 + r) \geq -1$, so the left-hand inequality can never be broken and there cannot be a period-doubling bifurcation.

The right-hand inequality can be broken in two ways. The point where $f_x(x_1) = -2$ can indicate the point where the period-two solution is created from a fixed point, or in some examples it can represent a saddle–node bifurcation, where two period-two cycles merge and no longer exist beyond the bifurcation point. The bifurcation at $f_x(x_1) = 0$ is known as a *symmetry-breaking bifurcation*. At this bifurcation, the symmetric period-two cycle becomes unstable and two asymmetric period-two solutions are created. A bifurcation diagram for this scenario is shown in Fig. 7.5.

In summary, for $x_{n+1} - x_n = f(x_n, \mu)$ with $f(-x, \mu) = -f(x, \mu)$,

- $x = 0$ is a fixed point for all μ.
- The fixed point $x = 0$ can have a pitchfork or period-doubling bifurcation.
- A symmetrical period-two solution branching from $x = 0$:

    - cannot have a further period-doubling bifurcation;

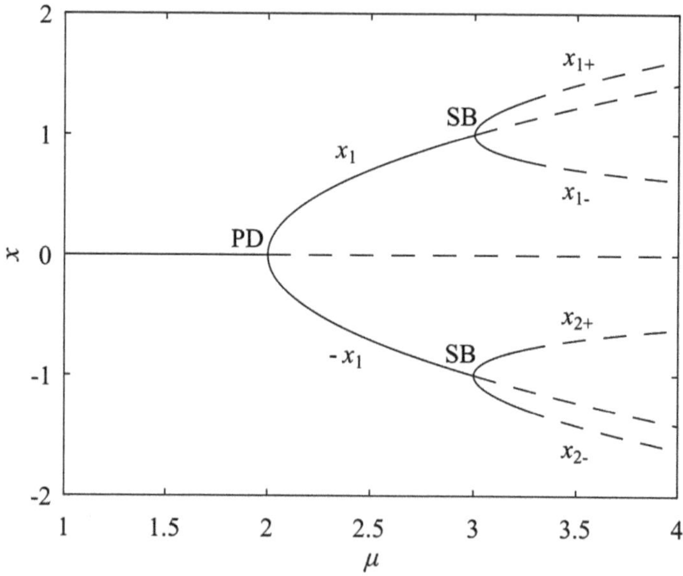

Fig. 7.5 Bifurcation diagram showing a symmetry-breaking bifurcation. The fixed point $x = 0$ has a period-doubling bifurcation (PD) leading to a symmetric period-two solution $(x_1, -x_1)$. When this becomes unstable at a symmetry-breaking bifurcation (SB), two asymmetric period-two solutions are created, labelled (x_{1+}, x_{2+}) and (x_{1-}, x_{2-}). The diagram is for the difference equation (7.28)

- can have a saddle–node bifurcation;
- can have a symmetry-breaking bifurcation creating two asymmetric period-two cycles.

Example 7.9 Find the period-doubling bifurcation, the symmetric period-two solution, and the symmetry-breaking bifurcation for

$$x_{n+1} - x_n = f(x_n, \mu) = -\mu x_n + x_n^3. \tag{7.28}$$

Then find the asymmetric period-two solution and determine when it is stable, and sketch the bifurcation diagram.

In this example, $x = 0$ is a fixed point for all μ, as required by the symmetry $x_n \leftrightarrow -x_n$. Also, $f_x(0, \mu) = -\mu$, so $x = 0$ is stable for $0 < \mu < 2$, with a pitchfork bifurcation at $\mu = 0$ and a period-doubling bifurcation at $\mu = 2$. A symmetric period-two solution of the form $(x_1, -x_1)$ has $x_1^2 = \mu - 2$, so this solution exists for all $\mu > 2$. On this period-two solution, $f_x = -\mu + 3x^2 = 2\mu - 6$, and this solution is stable if $(2\mu - 6)(2\mu - 4) < 0$, using (7.27), so $2 < \mu < 3$. There is a symmetry-breaking bifurcation at $\mu = 3$, where $x_1 = \pm 1$. Now to find the asymmetric period-two solution (x_1, x_2) that branches from the symmetry-breaking bifurcation, we need to solve

$$x_2 - x_1 = -\mu x_1 + x_1^3, \qquad x_1 - x_2 = -\mu x_2 + x_2^3. \tag{7.29}$$

Subtracting these two equations and removing the non-zero factor $(x_2 - x_1)$ as in the previous examples, gives[3]

$$2 = \mu - x_1^2 - x_1 x_2 - x_2^2. \tag{7.30}$$

We can also add the two equations in (7.29) and cancel a factor $(x_2 + x_1)$, since we are looking for a solution with $x_2 \neq -x_1$, and this leads to

$$0 = -\mu + x_1^2 - x_1 x_2 + x_2^2. \tag{7.31}$$

Adding (7.30) and (7.31) shows that

$$2 = -2x_1 x_2,$$

so $x_1 x_2 = -1$. This can now be used to eliminate x_2 from (7.31) to show that x_1 obeys

$$x_1^4 + (1 - \mu)x_1^2 + 1 = 0.$$

---

[3] This step and the following one use $x_2^3 - x_1^3 = (x_2 - x_1)(x_2^2 + x_1 x_2 + x_1^2)$ and $x_2^3 + x_1^3 = (x_2 + x_1)(x_2^2 - x_1 x_2 + x_1^2)$.

This has four real solutions for $\mu > 3$, so there are two asymmetric period-two solutions, related to each other by the symmetry. Finally, the stability of either of these asymmetric period-two solutions depends on

$$
\begin{aligned}
f_x(x_1) f_x(x_2) + f_x(x_1) + f_x(x_2) &= (-\mu + 3x_1^2)(-\mu + 3x_2^2) - \mu + 3x_1^2 - \mu + 3x_2^2 \\
&= \mu^2 - 2\mu + (3 - 3\mu)(x_1^2 + x_2^2) + 9x_1^2 x_2^2 \\
&= \mu^2 - 2\mu + (3 - 3\mu)(\mu - 1) + 9 \\
&= -2\mu^2 + 4\mu + 6,
\end{aligned}
$$

using the above equations for $x_1 x_2$ and $x_1^2 + x_2^2$. Hence these period-two states have a further period-doubling when this quantity is -2, which is at $\mu = 1 + \sqrt{5} \approx 3.236$. The resulting bifurcation diagram is shown in Fig. 7.5.

Example 7.10 Suppose that Newton's method is used to find the roots of $G(x) = x^5 - x^3 + \mu x$, where $\mu > 1/2$. Find the values of μ for which a period-two solution of Newton's method exists, and show that this period-two state can be stable.

The equation $G(x) = x^5 - x^3 + \mu x = 0$ has a solution $x = 0$. Other roots exist if $1 - 4\mu > 0$, but since $\mu > 1/2$ this is not true so there are no other roots. Newton's method (7.4) for $G(x) = 0$ is

$$
x_{n+1} = x_n - \frac{x_n^5 - x_n^3 + \mu x_n}{5x_n^4 - 3x_n^2 + \mu}, \tag{7.32}
$$

which is an odd function of x_n, of the type considered in this section. Suppose that there is a symmetric period-two cycle, $(x_1, -x_1)$. Then setting $x_{n+1} = -x_n = x_1$ in (7.32), rearranging, and factoring out the known solution $x_1 = 0$ gives a quadratic equation for x_1^2,

$$
9x_1^4 - 5x_1^2 + \mu = 0. \tag{7.33}
$$

This has two roots for x_1^2, corresponding to two symmetric period-two cycles, if $\mu < 25/36 = 0.6944\ldots$, so at this value of μ there is a saddle–node bifurcation of symmetric period-two cycles.

The stability of the period-two cycle can be found using the condition (7.27), $-2 < 2f_x(x_1) + f_x(x_1)^2 < 0$, where $f(x) = -G(x)/G'(x)$ and $f_x(x) = G(x)G''(x)/G'(x)^2 - 1$, see Example 7.3. If we choose a point where $G''(x) = 0$, then $f_x(x) = -1$ and so the cycle is superstable. This occurs when $x_1^2 = 3/10$ which satisfies (7.33) if $\mu = 0.69$. Hence there is a window of μ values near $\mu = 0.69$ where the period-two cycle is stable. This is fairly close to the saddle–node bifurcation where the period-two cycle is created. In fact there is only a narrow region of μ values for which the period-two solution is stable. Since the period-two solution can be stable, there are some choices of μ and initial conditions x_0 for which Newton's method fails to find the root $x = 0$, but instead flips between the two points on this cycle, as shown in Fig. 7.6.

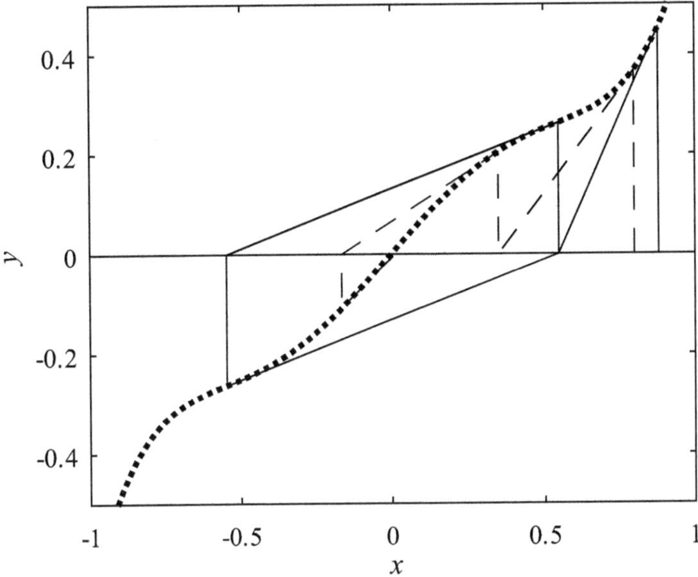

Fig. 7.6 Newton's method (7.32) applied to $G(x) = x^5 - x^3 + 0.69x = 0$. The dotted curve is $G(x)$. The dashed lines show the method with starting value $x_0 = 0.8$, which converges quickly to the root $x = 0$. The solid lines, for $x_0 = 0.88$, show the method converging to a period-two cycle

7.9 Period-Doubling Bifurcations of Periodic Orbits

As described in Sect. 7.3, investigating the stability of a periodic orbit by looking at points where nearby trajectories cross a line or plane known as a Poincaré section leads to a difference equation giving the difference between successive crossings of the section. Since difference equations can have period-doubling bifurcations, so can periodic orbits. In fact, this is the reason for the term "period-doubling".

When a stable periodic orbit has a supercritical period-doubling bifurcation, the orbit becomes unstable, and a stable, nearby periodic orbit is created. This new orbit has twice the period of the original one, and has two loops, as shown in Fig. 7.7. In the figure it appears that the trajectory crosses itself, but this is just because the figure shows a two-dimensional projection of a third-order system. In a second-order system, a periodic orbit cannot have a period-doubling bifurcation.

The system shown in Fig. 7.7 is

$$\dddot{x} = -x - a\dot{x} - b\ddot{x} + x^2, \tag{7.34}$$

which has fixed points at $x = 0$ and $x = 1$. The fixed point at $x = 0$ is stable for $ab > 1$ and has a supercritical Hopf bifurcation at $ab = 1$, so a stable periodic orbit exists for $ab < 1$ near this bifurcation.

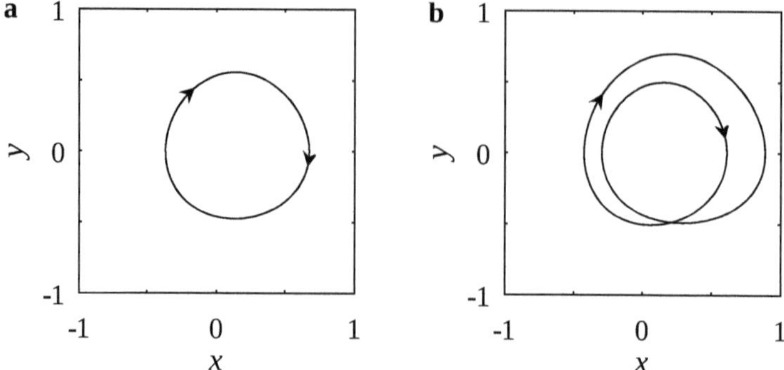

Fig. 7.7 Two-dimensional projection of the phase portrait of (7.34), using the variables x and $y = \dot{x}$. (**a**) For $a = 1$, $b = 0.7$, there is a stable periodic orbit. (**b**) At $a = 1$, $b = 0.6$, an orbit similar to the one shown in (**a**) exists but is unstable, and the stable solution is a period-two orbit

Numerical simulations of (7.34) show that for $a = 1$, the periodic orbit is stable as b is reduced below the bifurcation point $b = 1$ until $b \approx 0.63$, where a period-doubling bifurcation occurs, as shown in Fig. 7.7. Period-doubling bifurcations of periodic orbits are quite common in nonlinear differential equations of order three or higher.

7.9.1 Symmetric Periodic Orbits

The results of Sect. 7.8 have an important consequence for periodic orbits: *a symmetric periodic orbit cannot have a period-doubling bifurcation.*

Suppose that a system of $n > 2$ differential equations has a symmetry under a sign change of two or more of the variables. For example, if the nonlinear term in (7.34) is changed to x^3, then the third-order equation can be written as the three first-order equations

$$\dot{x} = y, \quad \dot{y} = z, \quad \dot{z} = -x - ay - bz + x^3, \tag{7.35}$$

with the symmetry of a sign change of each variable, $(x, y, z) \rightarrow (-x, -y, -z)$. Since the linear terms are the same as those in (7.34), the fixed point at the origin has a Hopf bifurcation when $ab = 1$, creating a periodic orbit that has this symmetry. More generally, a periodic orbit in (7.35) may or may not possess this symmetry, just as period-two solutions of (7.28) may or may not be symmetrical.

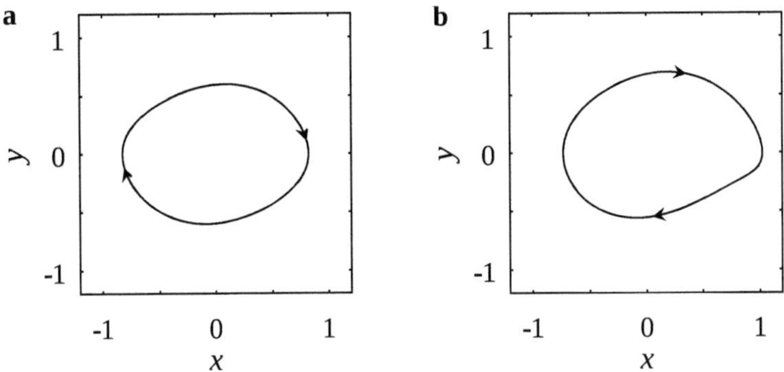

Fig. 7.8 Projection of the phase portrait of (7.35) onto the (x, y) plane, showing a symmetry-breaking bifurcation. (**a**) For $a = 0.5$, $b = 1$, there is a stable periodic orbit with a π-rotation symmetry. (**b**) At $a = 0.5$, $b = 0.8$, the symmetrical orbit is unstable, and an asymmetric periodic orbit is stable

Consider a symmetrical orbit in (7.35) projected onto any of the three two-dimensional coordinate planes, for example the orbit in the (x, y) plane shown in Fig. 7.8a, with the symmetry $(x, y) \rightarrow (-x, -y)$ of a rotation through an angle π in this plane. Now construct a Poincaré section by choosing the line $y = 0$. As solution trajectories cross this line, successive crossings (including those with $\dot{y} > 0$ and $\dot{y} < 0$) generate a difference equation relating the values of x at each crossing. Furthermore, the π-rotation symmetry guarantees that this difference equation has the $x \leftrightarrow -x$ symmetry investigated in Sect. 7.8, and the symmetrical periodic orbit corresponds to a symmetrical period-two solution $(x_1, -x_1)$ of this difference equation. Applying the main result of Sect. 7.8, there can be no period-doubling from the period-two solution of the difference equation, and hence no period-doubling of the periodic orbit. A period-doubling cannot be hidden by the projection, because the same argument can be applied to a different projection. A thorough study of this constraint was given by Swift and Wiesenfeld in their 1984 paper, *Suppression of Period Doubling in Symmetric Systems*.

Figure 7.8 shows a stable periodic orbit of (7.35) for $a = 0.5$, $b = 1$, which has a rotation symmetry, and one for $a = 0.5$, $b = 0.8$, which does not have this symmetry. The symmetrical periodic orbit still exists, but it is unstable, and two asymmetric periodic orbits have been created at a symmetry-breaking bifurcation. Only one of these is shown in Fig. 7.8b; the other one is obtained by rotating the phase portrait through an angle π. The behaviour of the differential equations (7.35) is analogous to the behaviour of difference equations such as (7.28).

Key Points from Chap. 7

- Difference equations of the form $x_{n+1} - x_n = f(x_n)$ can be used to model processes that change over finite time intervals. They can be thought of as the discrete analog of first-order differential equations.
- Difference equations arise when differential equations are discretised in order to find an approximate numerical solution on a computer. There are many ways to do this, but the simplest is known as Euler's method.
- Difference equations also occur when studying the stability of a periodic orbit, by monitoring the discrete points at which the periodic orbit crosses a line or plane known as a Poincaré section.
- A difference equation $x_{n+1} - x_n = f(x_n)$ has a fixed point at $x = x_*$ if $f(x_*) = 0$, as for a first-order differential equation.
- A fixed point x_* of a difference equation is stable if $f'(x_*) < 0$ (as for a differential equation) and $f'(x_*) > -2$. This second stability limit has no equivalent for fixed points of a differential equation.
- For a difference equation with a parameter, $x_{n+1} - x_n = f(x_n, \mu)$, a fixed point x_* has a stationary bifurcation as $f_x(x_*, \mu)$ passes through zero. There are three main types of stationary bifurcation, as for differential equations: saddle–node, transcritical and pitchfork.
- As $f_x(x_*, \mu)$ passes through -2, the fixed point has a *flip* or *period-doubling* bifurcation, where the fixed point becomes unstable and a new state is created in which the solution jumps to and fro between two values. Like a pitchfork bifurcation, a period-doubling bifurcation can be supercritical, if the period-two state is stable, or subcritical, if it is unstable.
- A period-two state can have a further period-doubling, leading to a period-four solution. However, if the system and its period-two solution have a symmetry $x_n \leftrightarrow -x_n$, this cannot occur. Instead, a symmetric period-two state can have a *symmetry-breaking bifurcation*, creating two asymmetric period-two states.
- The bifurcations discussed here for difference equations, such as period-doubling and symmetry-breaking bifurcations, can also occur for periodic solutions of differential equations.

Exercises

7.1 (a) Find the general solution of Fibonacci's difference equation (7.2) by trying a solution of the form $x_n = Ap^n$ for some real number p. Hint: it may be useful to recall the method for a second-order linear differential equation, see Sect. 2.4.

(b) Find the specific solution that obeys the initial conditions $x_0 = x_1 = 1$, and check that your formula gives the correct value for x_2.

(c) What is the limit of the ratio x_{n+1}/x_n as $n \to \infty$? What is the name of this ratio?

7.2 Write down Newton's method for the solution of $\sin x = 0$.

(a) What is the order of convergence of the method for this problem?

(b) For the initial guess $x_0 = 3$, how close to π is x_2? How does this relate to (a)?

(c) What happens if $x_0 = 1.5$?

(d) Which of the roots can be found by choosing different initial values in the range $0 < x_0 < \pi$?

(e) Is this a useful method for evaluating π accurately?

7.3 Halley's iterative method[4] for solving $G(x) = 0$ is

$$x_{n+1} = x_n - \frac{2G(x_n)G'(x_n)}{2G'(x_n)^2 - G(x_n)G''(x_n)}.$$

(a) Use two steps of Halley's method to approximate a root of the quadratic equation derived in Exercise 7.1, using an initial guess $x_0 = 2$.

(b) Apply two steps of Newton's method from the same initial value and compare the results. Which method converges faster?

7.4 Using your favourite programming language, code up the Euler's method pseudocode example for $\dot{x} = x - x^2$ with $x(0) = 2$ from the end of Sect. 7.2. Investigate the effect of increasing the step size. What is the largest step size h for which the method approaches 1 as the number of steps increases? Verify your numerical result analytically by finding the first step of the numerical method and considering the fixed points of the difference equation.

7.5 (a) Using the methods of Chap. 2, find the solution of $\dot{x} = x - x^2$ with $x(0) = 2$. Hint: use partial fractions.

(b) Adapt your code from Exercise 7.4 so that it displays the exact solution, the numerical solution and the error at each time step, for $h = 0.2$. At which time step is the error largest?

---

[4] Edmond Halley, 1656–1742, best known for predicting the return of the comet that bears his name.

7.6 (a) Using polar coordinates, construct the difference equation for successive crossings of the positive x axis for the system

$$\dot{x} = x - 2y - x(x^2 + y^2), \qquad \dot{y} = 2x + y - y(x^2 + y^2).$$

Hint: make use of Example 5.11 and Exercise 3.3.

(b) Check that the behaviour of the difference equation agrees with what you would expect, based on similar systems in Chap. 5.

7.7 Find all the fixed points of the difference equation

$$x_{n+1} - x_n = x_n^4 + \frac{x_n^3}{2} - x_n^2 - \frac{x_n}{2}.$$

For each fixed point, investigate whether it is unstable, stable or superstable.

7.8 (a) Write down the difference equation that arises when the logistic equation $\dot{x} = x - x^2$ is discretised using Euler's method with a step size h. Show that the fixed points of the difference equation are the same as those of the differential equation, for all values h.

(b) Consider the stable fixed point of the logistic equation. Show that in the difference equation, this fixed point is stable for small h but has a period-doubling bifurcation at $h = 2$. Why was a different answer found in Exercise 7.4?

(c) By using the method of Example 7.5, show that the points that form the period-two solution obey the quadratic equation $h^2 x^2 - h(h + 2)x + h + 2 = 0$. Hence show that the period-doubling bifurcation is supercritical. Show that if $h = 5/2$, one of the points of the period-two solution is $3/5$ and find the other.

(d) Adapt your code from Exercise 7.4 so that the initial condition is fairly close to the stable fixed point. Check that your code shows a period-doubling bifurcation at $h = 2$ by running it with h just below 2 and just above 2.

(e) Does your code give the two values found in (c) when $h = 2.5$? Try different initial conditions and explain the results.

7.9 Find the stationary bifurcations of the difference equation $x_{n+1} - x_n = f(x_n, \mu)$ for each of the following forms of $f(x, \mu)$, and determine their type.

(a) $f(x, \mu) = e^x - x + \mu$,
(b) $f(x, \mu) = x/(1 + x^2) - \mu x$.

7.10 For each of the following functions $f(x, \mu)$, find the location of any period-doubling bifurcations in the difference equation $x_{n+1} - x_n = f(x_n, \mu)$. If possible, find the period-two solution explicitly and hence find out whether the bifurcation is supercritical or subcritical.

(a) $f(x, \mu) = \mu x - x^2$.
(b) $f(x, \mu) = -\mu \log x$, for $x > 0$.
(c) $f(x, \mu) = x/(1 + x^2) - \mu x$.
(d) $f(x, \mu) = -\mu x/(1 + x^2)$.

7.11 (a) Write down the difference equation for Newton's method applied to the cubic $x^3 - x + \mu = 0$, $\mu > 0$.

(b) Show that a period-two cycle exists in which one of the points is $x = 0$, for a particular value of μ.

(c) Show that this period-two cycle is superstable. Hint: use Example 7.3 and (7.23).

7.12 (a) Write down the Newton iteration formula for solving $11x^4 - 6x^2 - 1 = 0$.

(b) Show that there are two possible symmetric period-two solutions $(x_1, -x_1)$.

(c) Show that one of these period-two solutions is stable and the other is stable.

(d) Check your results numerically by modifying your code from Exercise 7.11. What happens if the initial condition is set to one of the points of the unstable period-two solution?

7.13 The midpoint method for the numerical approximation of a solution to the differential equation $\dot{x} = f(x)$ with step size h is defined by the difference equation

$$X_{n+1} = X_n + hf\left(X_n + \frac{1}{2}hf(X_n)\right).$$ (7.36)

(a) Show that if x_* is a fixed point of the differential equation, then x_* is also a fixed point of the difference equation.

(b) Show that the converse is not true, by considering $f(x) = x(1 - x)$ and finding all the fixed points of the difference equation, including two 'spurious' fixed points that depend on h and are not fixed points of the differential equation.

(c) Investigate the stability of the fixed point at $X_* = 1$ in the difference equation, for $f(x) = x(1 - x)$. Compare this result with the case of Euler's method, studied in Exercise 7.8(b), giving one similarity, and one difference.

(d) What type of bifurcation occurs when the fixed point at $X_* = 1$ becomes unstable?

7.14 Consider the difference equation

$$X_{n+1} - X_n = -\mu \sin X_n, \quad \mu > 0.$$

(a) Find the fixed points, investigate their stability and find the location of any period-doubling bifurcations.

(b) Find the values of μ for which a symmetric period-two cycle $(x_1, -x_1)$ exists.

(c) Using the methods of Sect. 7.8, show that there is a symmetry-breaking bifurcation at $\mu = \pi$.

(d) Find the range of values of μ for which an asymmetric period-two cycle, (x_1, x_2), $x_2 \neq -x_1$, exists, and check that this is consistent with part (c).

(e) Show that there are infinitely many saddle–node bifurcations of symmetric period-two cycles, and derive their approximate location for large μ.

Chapter 8
Chaos

The ideas and methods introduced so far in this book come together in this final chapter on chaos. Chaos can occur in differential equations, where the variables are continuous functions of time, or in difference equations, where the system evolves in discrete steps. Chaotic systems show apparently irregular, nonperiodic behaviour, even though they can often be described by equations that look deceptively simple. The essential feature of chaos is that small changes to initial conditions grow exponentially, leading to large changes in the solution, though the trajectory remains in a bounded region of phase space. Chaos can often arise following a sequence of bifurcations that follow a remarkable universal pattern across different systems.

The first part of this chapter describes chaos in differential equations, concentrating on a famous third-order system. The second half of the chapter looks at chaos in difference equations. It turns out that this case is much easier to analyse, since chaos can occur in very simple first-order difference equations.

8.1 The Lorenz Equations

One of the most influential scientific papers of the last hundred years is *Deterministic Nonperiodic Flow*, written in 1963 by Ed Lorenz, a meteorologist at Massachusetts Institute of Technology (MIT). Lorenz was interested in the circulation of the Earth's atmosphere, and came up with a very simple model for the motion of a single convection cell, consisting of just three coupled differential equations. Using a computer to investigate this system, Lorenz found that it behaved in a very strange way. Although the motion was deterministic—uniquely specified by the differential equations and the initial conditions—solutions oscillated in a way that appeared almost random, rotating in one direction for a while and then reversing, and never exactly repeating itself. He also found that trajectories starting from two very close initial conditions would diverge from each other exponentially fast, and

© The Author(s), under exclusive license to Springer Nature Switzerland AG 2025
P. C. Matthews, *Differential Equations, Bifurcations and Chaos*,
Springer Undergraduate Mathematics Series,
https://doi.org/10.1007/978-3-031-99543-9_8

soon become quite different. Despite this, all the solutions somehow remained in the same finite region of the three-dimensional phase space. For some reason, Lorenz's term for this phenomenon, *deterministic nonperiodic flow*, did not really catch on, and it became known as *chaos*.

Lorenz's 1963 paper has been cited by other research papers more than 28,000 times. His harmless-looking equations are in fact so complicated that, for example, twenty years later, Colin Sparrow of the University of Cambridge wrote a book about them, running to over 250 pages.

8.1.1 Derivation of the Lorenz Equations

In his 1963 paper, Lorenz derived his equations from a severe truncation of the equations of fluid dynamics and heat transfer, which cannot be rigorously justified but captures the essential features of a convection cell. However, the equations can be derived quite simply for a fluid confined to a circular tube and heated from below, as shown in Fig. 8.1a. The warmer fluid at the bottom of the tube is less dense than the fluid at the top, causing the fluid to start to rotate either in a clockwise or anticlockwise direction.

In the nearby Mathematics Department of MIT, Willem Malkus, who had a fluid dynamics laboratory, realised that Lorenz's equations also described a waterwheel, where water was fed from above into buckets with a hole at the bottom, so that the water would slowly leak out, see Fig. 8.1b. Malkus designed and built the chaotic

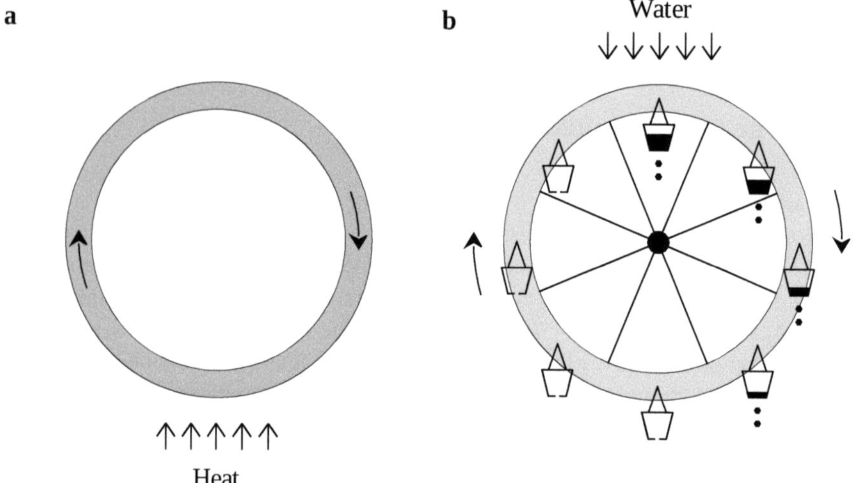

Fig. 8.1 The Lorenz equations can be thought of as a model for (**a**) the overturning of a fluid in a circular tube, heated from below, or (**b**) the motion of a waterwheel with leaky buckets, fed from above

waterwheel, and numerous examples have been built since then, to demonstrate the phenomenon of chaos.

The derivation given here is for the circular tube model. A fluid, such as water or air, is held within a circular tube, and heated from below. The hot fluid is buoyant and rises, causing it to rotate around the tube. Let θ be the angle of a point in the tube, measured anticlockwise from the bottom of the tube, and let v be the angular velocity (radians per second) of motion of the fluid in the anticlockwise direction. The actual velocity (metres per second) of the fluid motion is then Rv, where R is the radius of the circular tube. The fluid is assumed to be incompressible, so v does not depend on θ, only on time t. The temperature, $T(\theta, t)$, depends on both θ and t. The buoyancy force is proportional to the temperature, so a small section of fluid in the tube, in the interval $[\theta, \theta + \delta\theta]$ exerts an upward force $\alpha T(\theta)\delta\theta$, where α is a constant. The component of this force acting in the direction of increasing θ is $\alpha T(\theta, t) \sin\theta \, \delta\theta$. There is also a drag force due to the friction of the fluid passing the walls of the tube, which is proportional to v. The equation of motion for the rotating fluid is the angular form of Newton's second law, which is

$$I\dot{v} = \int_0^{2\pi} R\alpha T(\theta, t) \sin\theta \, d\theta - \beta v, \tag{8.1}$$

where β is a positive constant and I is the moment of inertia of the fluid in the tube. Assuming that the thickness of the tube is negligible compared with the radius R, I is the mass of the fluid multiplied by R^2.

Now consider how the temperature of the fluid changes. There are three effects to consider. The first is known as advection—in a time interval δt, the fluid that is at position θ at time t moves to $\theta + v\delta t$. Hence the change in temperature due to advection is $T(\theta + v\delta t, t + \delta t) = T(\theta, t)$, and using a Taylor expansion for small δt gives

$$\frac{\partial T}{\partial \theta} v + \frac{\partial T}{\partial t} = 0.$$

The second mechanism of temperature change is cooling due to heat loss to the surroundings, which is assumed to be proportional to the temperature, measured relative to the temperature outside the tube. This is known as Newton's law of cooling. The third is the heating applied to the system, constant in time, represented by a function $f(\theta)$ that is assumed to be an even function of θ. Adding these three effects together, the temperature equation is

$$\frac{\partial T}{\partial t} = -\frac{\partial T}{\partial \theta} v - kT + f(\theta), \tag{8.2}$$

where the constant k measures the cooling rate. Now since the temperature is a periodic function of θ, it can be written as an infinite Fourier series,

$$T(\theta, t) = a_0 + a_1 \cos\theta + b_1 \sin\theta + a_2 \cos 2\theta + b_2 \sin 2\theta + \ldots, \tag{8.3}$$

where the Fourier coefficients $a_0, a_1, b_1, a_2, b_2 \ldots$ are functions of t. At first glance it may appear that this leads to an infinite number of coupled differential equations for all the coefficients a_i and b_i, but this is not the case: in fact the result is just three equations, for v, a_1 and b_1. Substituting the Fourier series (8.3) into (8.1) gives

$$I\dot{v} = R\alpha\pi b_1 - \beta v. \tag{8.4}$$

This is because the trig integrals that appear are all zero except for the one arising from the $b_1 \sin\theta$ term (the trig functions are said to be *orthogonal* on the interval $[0, 2\pi]$, see Exercise 8.1).

To find an equation for b_1, we substitute the Fourier expansion (8.3) into (8.2) and pick out the terms in $\sin\theta$ (this is justified by the orthogonality of the trig functions), giving

$$\dot{b}_1 = a_1 v - k b_1. \tag{8.5}$$

The $f(\theta)$ term in (8.2) does not contribute to this equation, because $f(\theta)$ is an even periodic function, so its Fourier series only contains cosine terms, $f(\theta) = f_0 + f_1 \cos\theta + f_2 \cos 2\theta + \ldots$ with no $\sin\theta$ component. Since a_1 appears in (8.5), we need an equation for this, which comes from the $\cos\theta$ terms in (8.2):

$$\dot{a}_1 = -b_1 v - k a_1 + f_1. \tag{8.6}$$

The three equations (8.4), (8.5), and (8.6), are essentially the Lorenz equations. A rescaling in the form of a nondimensionalisation (see Example 3.4) turns them into the standard form. We replace the variables v, b_1 and a_1 by new dimensionless variables x, y and z defined by

$$v = kx, \quad b_1 = \frac{\beta k}{R\alpha\pi} y, \quad a_1 = \frac{f_1}{k} - \frac{\beta k}{R\alpha\pi} z,$$

and rescale time by a factor of k. After some simple algebra this leads to

$$\dot{x} = \sigma(y - x), \tag{8.7}$$

$$\dot{y} = rx - xz - y, \tag{8.8}$$

$$\dot{z} = xy - z. \tag{8.9}$$

These are the Lorenz equations. The two dimensionless parameters are defined by

$$\sigma = \frac{\beta}{kI}, \qquad r = \frac{f_1 \alpha\pi}{\beta k^2},$$

and they are both positive. The parameter σ measures the ratio of friction, which depends on the viscosity of the fluid, to the rate of heat loss, and is known as the

Prandtl number. The second parameter, r, is called the Rayleigh number in the context of fluid convection, and measures the ratio of the strength of the heating to the two damping parameters β and k. In Lorenz's paper there is a third parameter, appearing in front of the z term in (8.9), related to the choice of aspect ratio of the convection cell. In the model described here, this geometrical factor is fixed by the circular geometry and the coefficient is equal to 1. In this book we will use the two-parameter version of the equations, which makes some of the formulas slightly simpler.

The physical interpretation of the three dependent variables x, y and z is as follows. x represents the rotation rate of the fluid in the tube. y measures the left-right temperature difference, and z describes the up-down temperature difference, measured relative to the equilibrium state where a non-zero temperature difference is maintained by a balance between the applied heating and the heat loss.

The derivation of the Lorenz equations for the waterwheel model shown in Fig. 8.1b is almost exactly the same as the calculation above, see Exercise 8.2. The distribution of water around the buckets attached to the wheel plays the role of the fluid temperature.

8.1.2 Boundedness of the Lorenz Equations

An important property of the Lorenz equations is that solutions are bounded. This means that all trajectories of the system eventually enter a finite, bounded region of the three-dimensional phase space, and then remain within it. To demonstrate this, Lorenz considered the function

$$L(x, y, z) = x^2 + y^2 + (z - r - \sigma)^2.$$

Clearly $L(x, y, z) \geq 0$, and the surface $L(x, y, z) = \text{constant}$ is a sphere, centred at the point $x = y = 0, z = r + \sigma$. As the constant is increased, these surfaces form a sequence of concentric spheres. To show that solutions are bounded, consider the rate of change of L:

$$
\begin{aligned}
\dot{L}/2 &= x\dot{x} + y\dot{y} + (z - r - \sigma)\dot{z} \\
&= \sigma x(y - x) + y(rx - xz - y) + (z - r - \sigma)(xy - z) \\
&= -\sigma x^2 - y^2 - z^2 + z(r + \sigma) \\
&= -\sigma x^2 - y^2 - \left(z - \frac{r}{2} - \frac{\sigma}{2}\right)^2 + \frac{(r + \sigma)^2}{4}.
\end{aligned}
$$

The function L has been chosen so that several terms cancel out when finding $\dot{L}$. After using completion of the square in the last line above, we can see that for any choice of σ and r, $\dot{L} = 0$ on an ellipsoid, E, with its centre at $x = y = 0$, $z = (r + \sigma)/2$. Outside the ellipsoid E, $\dot{L} < 0$ and inside E, $\dot{L} > 0$.

Now consider the surface of a sphere, S, given by $L = $ constant, where the constant is chosen to be large enough that it contains the ellipsoid E. Then on the surface S, $\dot{L} < 0$, so any trajectory that starts on S must be directed into S, and no trajectory can leave the interior of S. Also, L must be a decreasing function of time for any trajectory that starts outside S, so all solutions must end up inside S.

8.1.3   Fixed Points and Their Stability

The fixed points of the Lorenz equations and their stability can be investigated in the same way as for the second-order systems discussed in Chap. 5. To do this we will need the Jacobian of the system, which is

$$J_L = \begin{pmatrix} -\sigma & \sigma & 0 \\ r - z & -1 & -x \\ y & x & -1 \end{pmatrix}. \tag{8.10}$$

It is also helpful to note that the Lorenz equations have a symmetry $x \leftrightarrow -x$, $y \leftrightarrow -y$, corresponding to the left-right symmetry of each of the systems shown in Fig. 8.1. In the x, y, z phase space, this sign change symmetry of x and y is a rotation through an angle π about the z axis. As discussed in Sect. 5.6, symmetries are often associated with invariant lines, and in this case the rotation symmetry means that the z axis, $x = y = 0$, is an invariant line of the Lorenz system.

Clearly the Lorenz system (8.7)–(8.9) has a fixed point at the origin, corresponding to no motion in Fig. 8.1. The stability of this point is determined by the eigenvalues λ of the Jacobian J_L at the origin, which obey the equation

$$(\lambda + 1)(\lambda^2 + (1 + \sigma)\lambda + \sigma - r\sigma) = 0.$$

All three eigenvalues are real. If $r < 1$, they are all negative, so the origin is stable, equivalent to a stable node in the second-order case. In this situation, the forcing of the system is not strong enough to overcome the effect of the drag force. But for $r > 1$ there are two negative eigenvalues and one positive one, so the origin is a kind of saddle point. Since one eigenvalue is positive, the origin is unstable. There is a stationary bifurcation at $r = 1$, and the symmetry of the system mentioned above suggests that this should be a pitchfork bifurcation.

If there are any other fixed points, then (8.7) shows that $y = x$, and then (8.9) implies that $z = x^2$. Then (8.8) gives $rx - x^3 - x = 0$, so apart from $x = y = 0$, there are two solutions,

$$x = y = \pm\sqrt{r - 1}, \qquad z = r - 1. \tag{8.11}$$

These two fixed points are related by the symmetry, and have a square root dependence on the bifurcation parameter, confirming that this is a pitchfork bifurcation. Also, since the new solutions exist for $r > 1$, where the origin is unstable, the pitchfork bifurcation is supercritical, so these fixed points should be stable near the bifurcation point. Physically, they correspond to a steady flow in which the water wheel or the fluid in the tube rotates at constant speed, either clockwise or anticlockwise.

Calculating the characteristic equation for these two symmetry-related fixed points (see Exercise 8.4) leads to

$$\lambda^3 + (\sigma + 2)\lambda^2 + (\sigma + r)\lambda + 2(r - 1)\sigma = 0. \tag{8.12}$$

It can be checked that, as expected, these fixed points are stable when r is just greater than 1. They could become unstable either at a stationary bifurcation, where $\lambda = 0$, or a Hopf bifurcation, when λ is purely imaginary. Seeking a stationary bifurcation by setting $\lambda = 0$ gives only the bifurcation at $r = 1$ that we already know about. Looking for a Hopf bifurcation, by setting $\lambda = i\omega$, where ω is real and non-zero, gives

$$- i\omega^3 - (\sigma + 2)\omega^2 + (\sigma + r)i\omega + 2(r - 1)\sigma = 0.$$

The real and imaginary parts of this equation are

$$(\sigma + 2)\omega^2 = 2(r - 1)\sigma, \qquad \omega^2 = \sigma + r \tag{8.13}$$

and eliminating ω^2 shows that r at the Hopf bifurcation is

$$r_H = \frac{\sigma(\sigma + 4)}{\sigma - 2}. \tag{8.14}$$

There are two possibilities, depending on the value of σ. If $\sigma < 2$, there is no Hopf bifurcation, and the fixed points (8.11) are stable for all $r > 1$. But for $\sigma > 2$, these fixed points are only stable for $1 < r < r_H$. For $r > r_H$ the fixed points have one real, negative eigenvalue, and a complex conjugate pair of eigenvalues with a positive real part, like an unstable spiral. At $r = r_H$, a periodic orbit is created. A gruesome calculation can be carried out to show that the Hopf bifurcation at $r = r_H$ is subcritical, so the periodic orbit is unstable, and exists for $r < r_H$.

This is as far as we can go using the methods discussed so far in this book. To investigate solutions for $r > r_H$, numerical methods must be used.

8.1.4 Chaos in the Lorenz Equations

A simple pseudocode for exploring the behaviour of the Lorenz equations using
Euler's method (7.7) is given below. The reader is strongly encouraged to write this
code in any suitable programming language. Investigating solutions numerically is
the best way to understand the remarkable complexity of these apparently simple
equations.

Pseudocode for Lorenz Equations

```
# Euler's method for the Lorenz equations
h = 0.02 # step size
nsteps = 1000 # number of steps
array t[0:nsteps], x[0:nsteps], y[0:nsteps], z[0:nsteps]
sigma = 5; r = 20; # set parameters
t[0] = 0; x[0] = 3; y(0) = 4; z(0) = 10 # initial conditions
for n from 0 to nsteps-1
    t[n+1] = t[n] + h
    x[n+1] = x[n] + h*sigma*(y[n] - x[n])
    y[n+1] = y[n] + h*(r*x[n] - x[n]*z[n] - y[n])
    z[n+1] = z[n] + h*(x[n]*y[n] - z[n])
end for
plot(t,x) # plot x(t)
plot(x,y) # plot phase space projected onto x,y plane
```

There are two parameters in (8.7)–(8.9), σ and r. Most investigations follow
Lorenz in setting $\sigma = 10$, but there is no particular reason for this choice. In this
section we will fix $\sigma = 5$ and allow r to vary. From the previous section, we know
that the origin is only stable for $r < 1$. There is a pitchfork bifurcation at $r = 1$ and
for $1 < r < r_H = 15$ the two fixed points (8.11) are stable. At $r = r_H = 15$ there
is a subcritical Hopf bifurcation.

Figure 8.2 shows $x(t)$ and $z(t)$ for a typical solution with $r = 17$. The variables
oscillate around the unstable fixed points at $x = y = \pm 4$, $z = 16$, but the
oscillations are irregular and nonperiodic, as described in the title of Lorenz's
paper. These irregular oscillations that look similar but never exactly repeat are
a characteristic feature of *chaos*, and the behaviour of the system is described as
chaotic.

Figure 8.3a shows $x(t)$ for two solutions, both with $\sigma = 5$ and $r = 25$. The initial
conditions for the two simulations differ only by a small change of 10^{-12} to each of
x, y and z. The two solutions $x_1(t)$ and $x_2(t)$ follow each other closely at first, but
separate at around $t = 60$ and are then quite different. This is known as *sensitive
dependence on initial conditions*. A very small change to the initial condition soon
leads to a large change in the solution trajectory. This is the most important defining
characteristic of chaos. To show the divergence between the two solutions more
clearly, Fig. 8.3b shows $\log(|x_1(t) - x_2(t)|)$. From the start of the simulation until

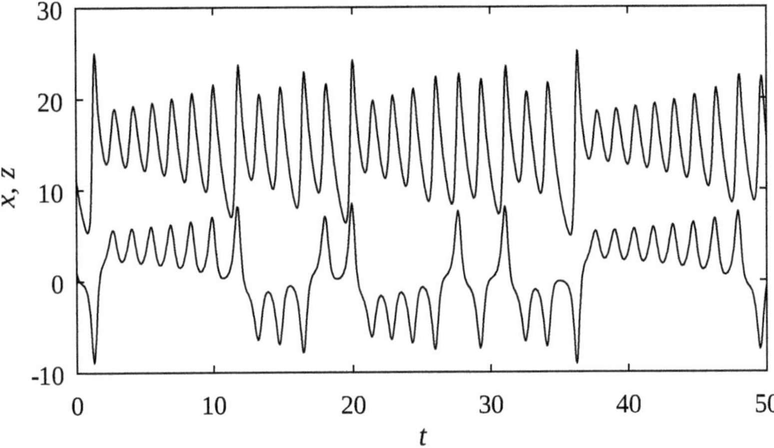

Fig. 8.2 $x(t)$ (lower curve) and $z(t)$ (upper curve) for the Lorenz equations with $\sigma = 5, r = 17$

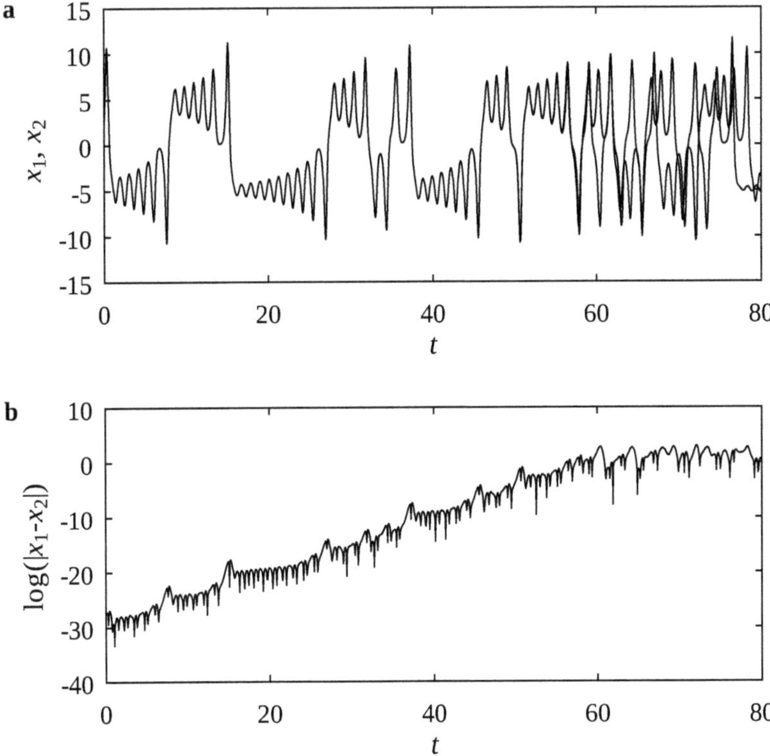

Fig. 8.3 (**a**) Plot of $x(t)$ for two simulations for $\sigma = 5, r = 25$, with initial condition differing by 10^{-12}. (**b**) The logarithm of the magnitude of the difference of the x values

the time at which the solutions visibly separate in the upper panel, this graph is approximately linear, with wiggles corresponding to the irregular oscillations. Since the logarithmic graph is roughly linear, the separation between the two solutions grows *exponentially* with time,

$$|x_1(t) - x_2(t)| \propto e^{\Lambda t}, \tag{8.15}$$

where the growth rate Λ is known as the *Lyapunov exponent*. The exponential growth of the separation between the two solutions stops when the separation is no longer small, as it must do, because the solutions are bounded by an ellipsoid, as shown in Sect. 8.1.2. It is difficult to measure the Lyapunov exponent, partly because the graph fluctuates around the straight line, and partly because we can only follow the exponential growth of the separation for a finite time. This time is limited on the right by the ellipsoidal bound, and on the left by machine precision (most computer systems cannot distinguish between two points with a separation below about 10^{-16}). However, we can estimate from the slope of Fig. 8.3b that $\Lambda \approx 0.5$, which is a typical value for the Lorenz equations. The value of Λ measures how rapidly trajectories separate, or 'how chaotic' the system is. A necessary condition for a system to be chaotic is that it must have a positive Lyapunov exponent. In fact, there are three Lyapunov exponents, similar in some ways to the three eigenvalues of a fixed point. The one which can be estimated from Fig. 8.3 is the largest of the three. It can be shown that one Lyapunov exponent is zero, and the sum of all three is the trace of the Jacobian matrix.

Remark 8.1 The sensitive dependence on initial conditions seen in the Lorenz equations and other chaotic systems has important consequences. It means that the future behaviour of a chaotic system is very difficult to predict. Any small error in the initial conditions is amplified exponentially with time. Similarly, the inevitable numerical errors of any numerical method, however small, grow exponentially (see Exercise 8.7(d)). The timescale over which the system can be predicted reliably is inversely proportional to the Lyapunov exponent. A good example of this is weather prediction. Despite the phenomenal increase in computer power over the years since Lorenz's paper was published, weather forecasts are still only accurate over a time period of a few days. Assuming that the Earth's weather is chaotic, which seems almost certain, since a hugely simplified model of it is chaotic, long-term weather forecasting is impossible.

Solutions of the Lorenz equations can also be represented as phase plane diagrams. Since the phase space is three-dimensional, we have to choose a two-dimensional projection. Figure 8.4 shows typical x, y and x, z phase planes for $\sigma = 5, r = 17$. The trajectories appear to intersect each other, but this is just a result of the projection; in fact one trajectory is above or below the other at each crossing. Clearly this cannot happen in a second order-system. Chaos can only occur in differential equations of order three or higher. The rotation symmetry of the system can be seen in the phase plane diagrams (the apparent reflection symmetry

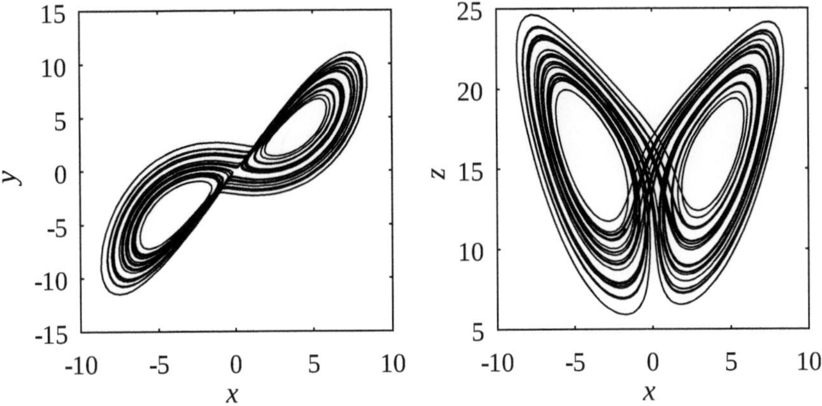

Fig. 8.4 Phase portraits in the x, y and x, z planes for the Lorenz equations with $\sigma = 5$, $r = 17$

in the x, z plane is again a result of the projection). The two 'holes' in the phase portraits contain the unstable fixed points at $(4, 4, 16)$ and $(-4, -4, 16)$.

The remarkable three-dimensional object shown in Fig. 8.4 adorns the covers of many books about chaos. It is known as a *strange attractor*. The word *strange* needs no explanation. An *attractor* is, as its name suggests, any object in phase space that attracts nearby trajectories. For example, stable nodes and stable periodic orbits are also attractors.

A typical trajectory shown in Fig. 8.4 spirals outward from one of the non-zero fixed points, moving away slowly since r is only slightly greater than r_H. After some time, the trajectory jumps across towards the other unstable fixed point, and then slowly spirals out from that one, and so on. In the spiralling phase, the trajectories appear to be almost confined to a flat disc. This is because when the parameter values are close to the Hopf bifurcation and the trajectory is close to the fixed point, the order of the system can be reduced as described in Sect. 6.2.1. The eigenvalues of the fixed point $(4, 4, 16)$ for $r = 17$, $\sigma = 5$ are approximately $\lambda_1 \approx -7.08$, $\lambda_{2,3} \approx 0.0416 \pm 4.75\mathrm{i}$. The real part of the negative eigenvalue is very much greater (by a factor of over 100) than the real parts of the two complex eigenvalues. Therefore there is rapid exponential decay in the direction of the eigenvector corresponding to the negative eigenvalue, which squashes solutions down onto a two-dimensional surface. This is investigated in the following example.

Example 8.1 For $\sigma = 5$, find the eigenvalues and eigenvectors of the non-zero fixed points at the Hopf bifurcation. Hence find the equations of the planes in which the Hopf bifurcations take place for each of the fixed points, and the angle between these planes. Compute the attractor and plot it as viewed from a point on one of these planes.

For $\sigma = 5$, $r_H = 15$ and $\omega^2 = 20$, from (8.13). The characteristic equation (8.12) is

$$\lambda^3 + 7\lambda^2 + 20\lambda + 140 = 0.$$

Two of the eigenvalues are $\pm i\omega$, and the sum of the eigenvalues must be -7, so the third eigenvalue is -7. The Jacobian matrix (8.10) at the fixed point $(x_0, x_0, 14)$, where $x_0 = \sqrt{14}$, is

$$J_L = \begin{pmatrix} -5 & 5 & 0 \\ 1 & -1 & -x_0 \\ x_0 & x_0 & -1 \end{pmatrix}.$$

A straightforward calculation gives the eigenvectors

$$v_{1,2} = \begin{pmatrix} 5x_0 \\ (5 \pm i\omega)x_0 \\ \omega^2 \mp 6i\omega \end{pmatrix}, \qquad v_3 = \begin{pmatrix} 20 \\ -8 \\ -x_0 \end{pmatrix}.$$

The two complex conjugate eigenvectors v_1 and v_2 generate a plane. Two real vectors in this plane are $(v_1 + v_2)/2 = \mathrm{Re}(v_1)$ and $(v_1 - v_2)/(2i) = \mathrm{Im}(v_1)$, and a vector normal to the plane is the cross product of these two, so we need to compute

$$\mathbf{n} = \begin{pmatrix} 5x_0 \\ 5x_0 \\ \omega^2 \end{pmatrix} \times \begin{pmatrix} 0 \\ \omega x_0 \\ -6\omega \end{pmatrix} = \begin{pmatrix} -30x_0\omega - \omega^3 x_0 \\ 30\omega x_0 \\ 5\omega x_0^2 \end{pmatrix} = 5x_0\omega \begin{pmatrix} -10 \\ 6 \\ x_0 \end{pmatrix}.$$

The equation of the plane is therefore $-10x + 6y + \sqrt{14}z = $ constant, and the value of the constant must be $10\sqrt{14}$ for the plane to pass through the fixed point. Note that this plane is not perpendicular to the eigenvector of the negative eigenvalue. Figure 8.5 shows the attractor viewed from a point on this plane, showing the part of the attractor near one of the fixed points as a slightly curved disc seen edge-on.

Using the symmetry of the Lorenz attractor, the tangent plane of the other fixed point is found by changing the sign of x and y, so $10x - 6y + \sqrt{14}z = 10\sqrt{14}$. The

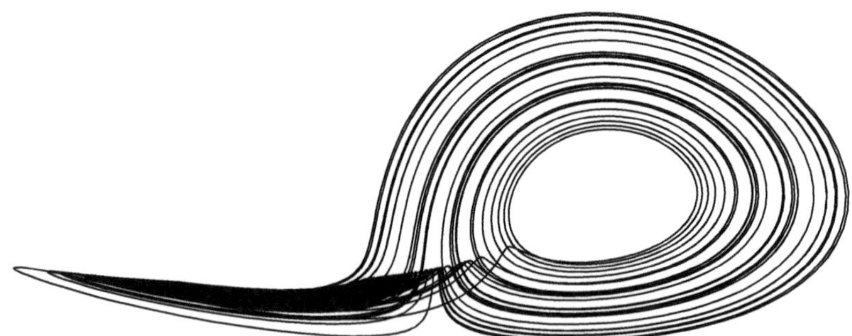

Fig. 8.5 The Lorenz attractor at the Hopf bifurcation, $\sigma = 5$, $r = 15$, viewed from a point in the plane of the Hopf bifurcation

angle α between the planes is the same as the angle between their normals, so using the dot product to find α,

$$(-10, 6, \sqrt{14}) \cdot (10, -6, \sqrt{14}) = (10^2 + 6^2 + 14)\cos\alpha,$$

giving $\cos\alpha = -122/150$, so $\alpha \approx 144°$, or $36°$ using the smaller angle at which the planes intersect.

As r is increased at fixed σ beyond the Hopf bifurcation, the chaotic behaviour persists for a large range of r values, and the appearance of the attractor remains similar. For $r = 40$, way beyond the Hopf bifurcation, the magnitude of the ratio of the real, negative eigenvalue to the real part of the complex ones is over 20, so the apparent structure of almost-flat spiralling discs around the fixed points remains.

However, as the parameter r is increased further, there are small windows of order in amongst the chaos. In these regions, stable periodic orbits are found. Figure 8.6a shows a stable periodic orbit that loops around one of the fixed points four times, and then around the other one four times, before repeating itself. Note that this periodic behaviour is similar to the chaotic dynamics, with solutions spiralling outward from first one fixed point and then the other. The orbit has the rotation symmetry $(x, y) \leftrightarrow (-x, -y)$ of the Lorenz equations. These types of periodic orbit can be described in a notation that indicates the way in which the orbit loops around the left and the right fixed point, so this orbit has the form LLLLRRRR. Around $r = 82$, there is a small interval where a stable orbit passes around one fixed point once and the other one twice, so its pattern is LRR, as shown in Fig. 8.6b. This orbit is not symmetrical, but the symmetry of the equations means that there is also a stable LLR periodic orbit with a phase portrait that is the same but rotated through π. Which of these orbits is found in a numerical simulation depends on the choice of initial condition.

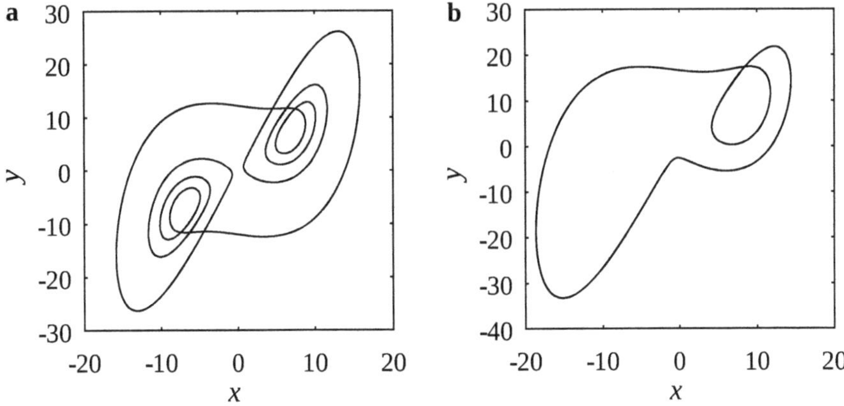

Fig. 8.6 Stable periodic orbits in the Lorenz equations. (**a**) $\sigma = 5, r = 55.5$; (**b**) $\sigma = 5, r = 82$

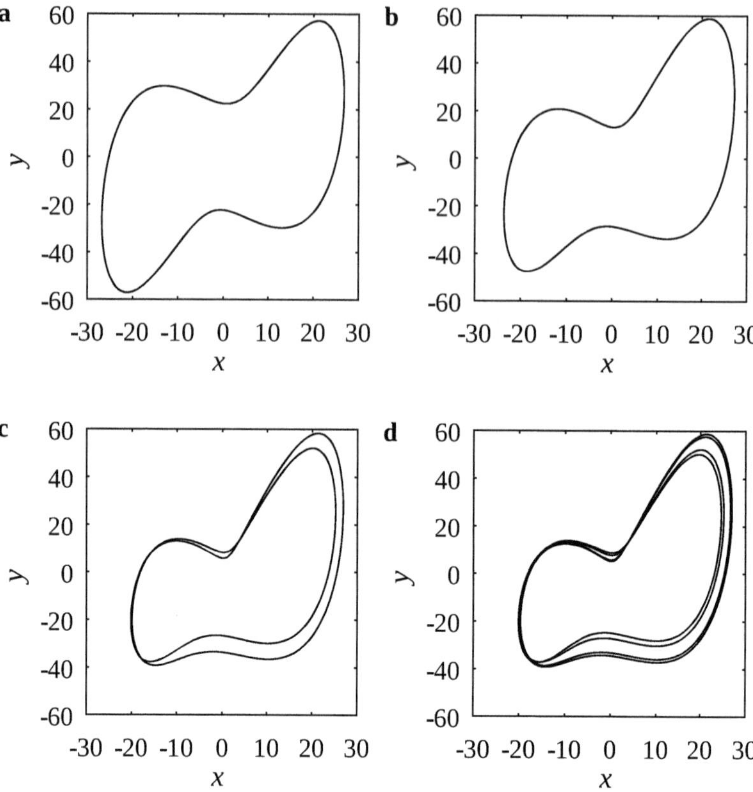

Fig. 8.7 Period doubling in the Lorenz equations, for $\sigma = 5$. (**a**) $r = 260$, symmetric orbit; (**b**) $r = 230$, asymmetric orbit; (**c**) $r = 181$, period-two orbit; (**d**) $r = 178$, period-four orbit

There are many more windows of stable periodic orbits for other values of r that can be explored (see Exercise 8.8). As r becomes larger, these windows get wider, until for very large r there is no chaotic attractor and a periodic orbit remains stable as $r \to \infty$. This behaviour is best illustrated by starting with a large value of r and then reducing r, monitoring any changes in behaviour—in other words, looking for bifurcations. Some of the stable periodic orbits found in this way are shown in Fig. 8.7. For $r = 260$ there is a symmetrical orbit that goes around both fixed points, so the pattern in the notation used above is LR, shown in part (a). Recall from Sect. 7.9.1 that this symmetrical orbit cannot have a period-doubling bifurcation. Somewhere near $r = 250$ this orbit becomes unstable at a symmetry-breaking bifurcation, where two stable asymmetric LR orbits are created, one of which is shown for $r = 230$ in part (b). As r is reduced further, this orbit becomes unstable at a period-doubling bifurcation near $r = 188$, creating a LRLR orbit with twice the period, shown in (c). This orbit in turn has a period-doubling bifurcation near $r = 180$, creating the LRLRLRLR orbit shown in part (d) for $r = 178$. For $r = 176$ the solution appears to be chaotic.

The behaviour shown in Fig. 8.7 is known as a *period-doubling sequence* or *period-doubling cascade*. As the parameter is varied, in this case decreased, an infinite sequence of period doubling bifurcations occurs, getting closer to each other and accumulating at a point. It becomes increasingly difficult to identify the bifurcation points, because they get very close to each other and near the bifurcation points the system evolves very slowly. Beyond this accumulation point the solution is chaotic. Hence this process is also known as the *period-doubling route to chaos*. This process occurs in many systems and will be discussed further in Sect. 8.3.

8.2 Features of Chaos

Although the Lorenz equations are the most well-known chaotic system, and the first in which chaos was identified and studied in detail, there is nothing particularly special or unique about them. There are many other systems of nonlinear differential equations that have chaotic behaviour. Examples of simple chaotic third-order systems include the Rössler equations,

$$\dot{x} = -y - z, \qquad \dot{y} = x + ay, \qquad \dot{z} = a + z(x - b),$$

the Genesio–Tesi equation,

$$\dddot{x} = -x - a\dot{x} - b\ddot{x} + x^2,$$

and the Rucklidge equations,

$$\ddot{x} + ax = b\dot{x} - xy, \qquad \dot{y} = -y + x^2,$$

where a and b are constants. Although they are written in different forms, each is equivalent to three coupled first-order differential equations. The first two examples have only one nonlinear term, and have no symmetry. The third has a symmetry like the Lorenz equations, as it is also derived from a model of heat transfer by convection. Each of these systems has a strange attractor for certain parameter values.

A necessary condition for chaos is sensitive dependence on initial conditions (SDIC), with trajectories diverging from each other exponentially. Another way of putting this is that there must be a positive Lyapunov exponent. But this alone is not sufficient, since the simple linear equation $\dot{x} = x$ has solutions that diverge from each other exponentially! To show chaos, a system must have SDIC but also must have bounded solutions. This condition of boundedness is satisfied by the Lorenz equations for all parameter values, as shown in Sect. 8.1.2, and by each of the three other systems given above for at least some values of the parameters.

This combination of properties may at first seem contradictory. How can solutions expand from each other exponentially, and yet remain within a bounded region of parameter space? The answer is that while expansion occurs in some directions, contraction occurs in other directions. For the Lorenz equations, the contraction dominates over the expansion. We can see this by considering the dynamics in phase space as the motion of a gas, with infinitely many molecules moving in three-dimensional space, defining a vector field representing the velocity of the gas. The expansion or contraction of a small blob of fluid is measured by the divergence of this vector field, which for the Lorenz equations is

$$\frac{\partial(\sigma y - \sigma x)}{\partial x} + \frac{\partial(rx - xz - y)}{\partial y} + \frac{\partial(xy - z)}{\partial z} = -\sigma - 2 < 0.$$

This is negative, so volume elements in phase space are constantly shrinking, and a consequence of this is that the volume of the Lorenz attractor is zero.

A useful concept to understand this process of expansion and contraction in a bounded domain is the idea of stretching and folding, shown in Fig. 8.8. Suppose that an object, for example a lump of dough used in bread-making, is stretched out by a factor of 2 in the x direction and compressed by a factor of 2 in the y direction. Then the right-hand half is folded over on top of the left half. The object then remains in the same region of space as its initial configuration. This process can then be repeated, creating a folded configuration with 4, then 8, then 16 layers, and so on. Only three stretch-fold cycles are shown in Fig. 8.8. This process is very similar to how kneading works in bread-making—the dough is repeatedly squashed down and spread out, and then folded over. The same process is used to make hand-pulled noodles. One significant difference between this model and the Lorenz attractor is that the stretching and folding of dough is volume-preserving, while volume elements shrink in the Lorenz equations, as shown above. We will return to this simple stretch-and-fold model in Sect. 8.3.

One more remarkable thing about chaotic systems concerns periodic orbits. We have seen that the Lorenz equations have many parameter windows where there is a stable periodic orbit, and that there are many different types of these periodic orbits. What happens to these orbits outside these windows, when the numerical simulations show chaos? The answer is that these periodic orbits usually still exist, but they are unstable. When there is a strange attractor, as in Fig. 8.5 for example, there are also infinitely many unstable periodic orbits. In fact, *periodic orbits are dense*. This means that at any point on the strange attractor, an unstable periodic orbit is nearby. More precisely, let P be any point on the strange attractor. Then for

Fig. 8.8 Schematic diagram of stretching and folding. An object is repeatedly stretched out by a factor of two and then folded over

any choice of $\varepsilon > 0$, there exists a point Q such that $|P - Q| < \varepsilon$ and Q lies on a periodic orbit. It is difficult to prove this for the Lorenz equations, but it can easily be shown for a particular type of difference equation, as shown in the next section.

8.3   Chaos in Difference Equations

Chaotic dynamics occurs in difference equations as well as in differential equations. The case of difference equations is in many ways much easier to investigate. For a system of differential equations to show chaos, the order of the system must be at least three, but chaos can occur in the first-order difference equation

$$x_{n+1} - x_n = f(x_n). \tag{8.16}$$

Section 8.3.1 below explores how a period-doubling sequence leading to chaos can occur in difference equations. The details of the dynamics of chaos are investigated in the later sections, where a number of results can easily be proved in some simple examples.

8.3.1   Period-Doubling Route to Chaos

In the previous chapter, we saw that a fixed point of a difference equation can become unstable at a period-doubling bifurcation, and that the period-two solution created at this bifurcation can itself have a period-doubling bifurcation, creating a period-four solution (see Example 7.6). Further investigation of the bifurcations in such systems can be carried out computationally, see the pseudocode example below.

```
Pseudocode for Difference Equations
    # Draw bifurcation diagram for x_n+1 = x_n + mu - x_n^2
    mumin=0.5; mumax=2.25; musteps=200; itmax=800
    for k from 1 to musteps
        mu[k] = mumin + (mumax - mumin)*(k - 1)/(musteps - 1)
        x[1,k] = 1.1
        for n from 1 to itmax
            x[n+1,k] = x[n,k] + mu[k] - x[n,k]^2
        end for
    end for
    plot(mu, x[itmax/2:itmax,1:musteps])
```

This code simulates the difference equation (7.19), $x_{n+1} - x_n = \mu - x_n^2$ used in Examples 7.5 and 7.6. The method is to take a large number of small steps in the

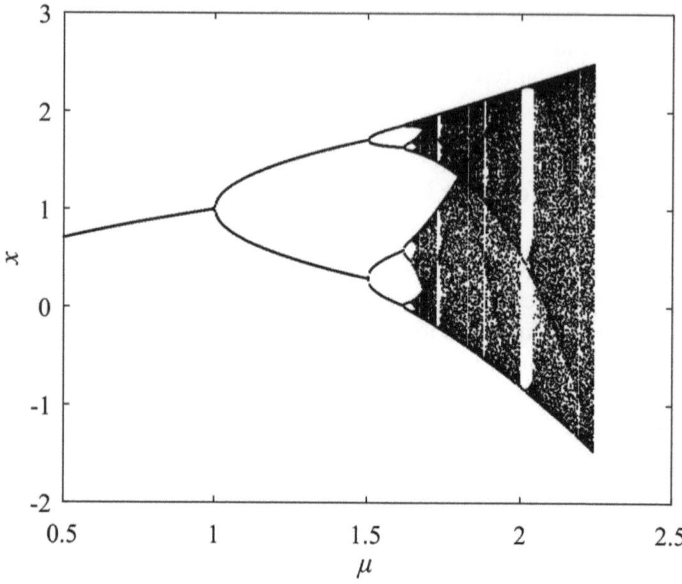

Fig. 8.9 Bifurcation diagram for $x_{n+1} - x_n = \mu - x_n^2$

parameter μ, between chosen minimum and maximum values. For each value of μ, an initial condition is chosen ($x_1 = 1.1$ in the code example), and a large number of iterations of the difference equation are carried out. The first half of the iterations are not used, since they include the transient solution that depends on the initial condition chosen. Each step of the second half of the iterations is plotted as a point, which gives the impression of a line if the number of steps in μ is large enough. Note that the exact form of the plot command is very language-dependent.

The result is shown in Fig. 8.9. For $\mu < 1$ there is a stable fixed point. A period-doubling bifurcation at $\mu = 1$ leads to a stable period-two cycle, which has a second period-doubling bifurcation at $\mu = 1.5$. This is all consistent with the results of Example 7.6. After this point there is another period-doubling bifurcation, and soon after that the diagram shows apparently scattered points, suggesting that the solution is chaotic. This is the same *period-doubling route to chaos* as was shown for the Lorenz equations in Fig. 8.7. Another similarity with the Lorenz equations is that there are narrow windows of regular periodic behaviour, for example there seems to be a region near $\mu = 2$ where a period-three solution is stable.

8.3.2   *Universality and Self-similarity*

Figure 8.10 shows the bifurcation diagram for $x_{n+1} - x_n = 2/x_n - \mu$, considered in Example 7.7. The striking thing is that although the numbers and the shapes of the

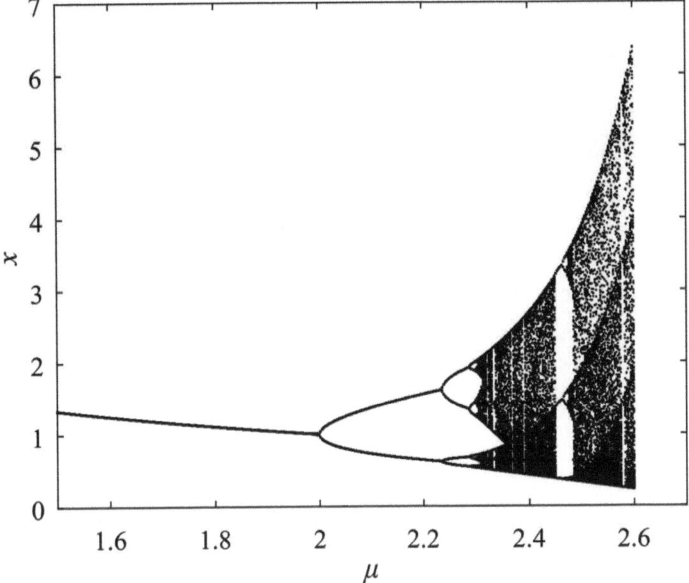

Fig. 8.10 Bifurcation diagram for $x_{n+1} - x_n = 2/x_n - \mu$

curves are different, the bifurcation diagram is remarkably similar to Fig. 8.9. There is the same sequence of accumulating period-doubling bifurcations, followed by a region of chaotic behaviour, and the same window of stable period-three solutions. In fact, there are many nonlinear difference equations that show qualitatively the same picture (see Exercise 8.11). This is known as *universality*.

Remarkably, this universality is not just qualitative. It is also quantitative, as discovered by Mitchell Feigenbaum in the 1970s. Feigenbaum looked carefully at the values of μ at the successive period-doubling bifurcations in various difference equations. He showed that the ratio of the gaps between the bifurcations obeyed a universal rule. Let μ_n be the value of μ at the nth period-doubling bifurcation. Then

$$\lim_{n \to \infty} \frac{\mu_{n+1} - \mu_n}{\mu_{n+2} - \mu_{n+1}} = \delta = 4.6692\ldots \tag{8.17}$$

for all systems that have a period-doubling cascade, under certain conditions. This law applies not just to difference equations, but also to differential equations such as the Lorenz system, and has even been seen in laboratory experiments.

Feigenbaum also found a second universal constant, relating the width of the branches emerging from each period-doubling bifurcation as the sequence progresses. These universal properties mean that the bifurcation diagrams show *self-similarity*: zooming in on a small region of the bifurcation diagram gives a picture very similar to the original one. Objects with this self-similar structure are often known as *fractals*.

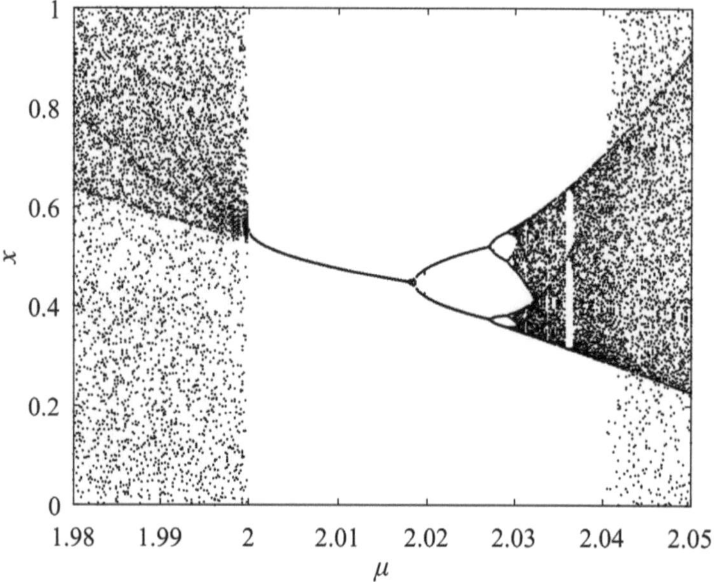

Fig. 8.11 Bifurcation diagram in the period-three window of $x_{n+1} - x_n = \mu - x_n^2$

Figure 8.11 shows an example of this self-similarity. The figure shows the same bifurcation diagram as Fig. 8.9, but looking closely at the period-three window near $\mu = 2$. Only the region $0 < x < 1$ is shown, including just one of the three points that make up the period-three cycle. The striking thing is that the sequence of bifurcations in this small window is very similar to the picture of the full system shown in Fig. 8.9. The period-three solution has a period-doubling bifurcation at $\mu \approx 2.018$, creating a period-six solution. There are then further period-doubling bifurcations to states of period 12, 24 ..., followed by a chaotic region and a small window where a period-nine orbit is stable. Similarly, focusing in on the even narrower period-nine window near $\mu = 2.036$ would produce another even smaller version of almost the same picture.

8.3.3   *Period Three and Sharkovsky's Theorem*

There is more to universality than period-doubling cascades. There are also some remarkable general results about difference equations that have period-three solutions. The first is that if (8.16) has a period-three solution, and the function f is continuous, then periodic solutions also exist of *any other period*. So in the period-three window near $\mu = 2$ in Fig. 8.9, although the stable behaviour is very simple, the full dynamics is extremely complicated. For any value of μ in this window, as well as the stable orbit of period three, there are also unstable cycles of period 7, 19 and 371, for example. Furthermore, as well as infinitely many periodic cycles, there are also unstable nonperiodic solutions, that is, chaotic ones. This result is generally referred to as *Period three implies chaos*, which was the title of a paper published in 1975 by Li and Yorke that showed this result (incidentally, this paper originated the term 'chaos').

Closely related to this result is Sharkovsky's theorem. Sharkovsky proved a more general result than Li and Yorke. He defined an ordering of integers in the following way. The Sharkovsky list of integers starts with powers of two: 1, 2, 4, 8, 16...and ends with all the odd numbers in reverse order, so the last numbers are 11, 9, 7, 5, 3. Before the odd numbers come all the odd numbers multiplied by 2: $2 \times 11, 2 \times 9,$ $2 \times 7, 2 \times 5, 2 \times 3$, and before this come $2^2 \times 11, 2^2 \times 9, 2^2 \times 7, 2^2 \times 5, 2^2 \times 3,$ and so on. So Sharkovsky's ordering is

$$1, 2, 4, 8, 16 \ldots 44, 36, 28, 20, 12 \ldots 22, 18, 14, 10, 6 \ldots 11, 9, 7, 5, 3$$

and every integer has its place in it. Sharkovsky's remarkable result is that for the difference equation (8.16) with f continuous, if there is an orbit of period n, then there is also an orbit of period m, for any m that comes before n in the sequence. So for $n = 3$, orbits of all other periods exist, and there is a similar result for any other integer. For example, suppose that there is an orbit of period 14. Then, according to Sharkovsky's theorem, there are also orbits of period 18, 22, 12, 20, 28 and any power of 2, but not necessarily orbits of period 10, 6, 11 or 9.

The proof of Sharkovsky's theorem is fairly long and involved. However, some results can be obtained easily, making use of the Intermediate Value Theorem (IVT). The IVT in the form needed here states that if $f(x)$ is a continuous function defined on the interval $[a, b]$, and $f(a)f(b) < 0$, so either $f(a) < 0$ and $f(b) > 0$, or $f(a) > 0$ and $f(b) < 0$, then there exists a point c with $a < c < b$ such that $f(c) = 0$. In words, if you draw a curve that starts on one side of a horizontal line and ends on the other side, without taking your pencil off the paper, then the curve must cross that horizontal line.

Example 8.2 If the function f is continuous, and the difference equation (8.16) has a period-two orbit, show that it has a fixed point.

Label the two points of the period-two cycle as a and b, with $a < b$. Then

$$b - a = f(a), \qquad a - b = f(b).$$

Since $a < b$, $f(a) > 0$ and $f(b) < 0$, exactly the conditions required for the IVT. So, by the IVT, there is a point c, with $a < c < b$ and $f(c) = 0$. The point c is then, by definition, a fixed point of the difference equation (8.16).

In view of the importance of period-three solutions, it is worth studying them more closely. At the left-hand edge of the window of period-three solutions shown in Fig. 8.11, the branch of the period-three solution becomes strongly curved and appears to have a square-root dependence on μ. This suggests that the period-three solution is created at a saddle–node bifurcation. Also, the period-three solution seems to start at precisely $\mu = 2$, which suggests that it should be possible to deduce this analytically!

There are various ways of finding the value of μ at which the period-three solution is created at a saddle–node bifurcation. They all involve a fair amount of algebra, but probably the most elegant method is the following, due to William Gordon.

Example 8.3 For $x_{n+1} - x_n = \mu - x_n^2$, derive analytically the range of values of μ for which a period-three solution exists, and deduce the type of bifurcation at which it is created.

Consider the formula

$$x_n = a + b\omega^n + \bar{b}\bar{\omega}^n, \tag{8.18}$$

where ω is the complex cube root of 1, $\omega = (-1 + \sqrt{3}i)/2$, and the bar denotes the complex conjugate. The following properties of ω will be useful:

$$\omega^3 = 1, \quad \omega^2 = \bar{\omega} = 1/\omega, \quad \omega + \bar{\omega} = -1. \tag{8.19}$$

If x_n has the form (8.18), then clearly $x_{n+3} = x_n$, so x_n has period three. It is also true, but not so obvious, that any period three cycle can be written in the form (8.18). Substituting (8.18) into the difference equation $x_{n+1} - x_n = \mu - x_n^2$,

$$b\omega\omega^n + \bar{b}\bar{\omega}\bar{\omega}^n - b\omega^n - \bar{b}\bar{\omega}^n = \mu - (a + b\omega^n + \bar{b}\bar{\omega}^n)^2.$$

Multiplying out the bracket gives nine terms, and the resulting equation contains constant terms, terms in ω^n and terms in $\bar{\omega}^n$. Since the equation must hold for all n, we can equate coefficients of each type of term, which gives the three equations

$$0 = \mu - a^2 - 2|b|^2, \tag{8.20}$$

$$b(\omega - 1) = -2ab - \bar{b}^2, \tag{8.21}$$

$$\bar{b}(\bar{\omega} - 1) = -2a\bar{b} - b^2. \tag{8.22}$$

Combining (8.21) and (8.22) shows that

$$|b|^2 = (\omega - 1 + 2a)(\bar{\omega} - 1 + 2a) = 3 - 6a + 4a^2,$$

using the properties (8.19). Finally, substituting this formula for $|b|^2$ into (8.20) gives a quadratic equation for a,

$$9a^2 - 12a + 6 - \mu = 0.$$

This has real solutions if $36 > 9(6 - \mu)$, so period-three solutions exist for $\mu > 2$, and there is a saddle–node bifurcation of period-three solutions at $\mu = 2$. For any given value of $\mu \geq 2$, it is then possible to solve for a and b and so find an explicit formula for the three values of x_n that form the period-three cycle (see Exercise 8.15).

8.3.4   Quantifying Chaos

Chaos in difference equations requires that solutions starting from two nearby initial conditions diverge from each other exponentially, a property known as sensitive dependence on initial conditions (SDIC). We will use the same informal definition of chaos in difference equations as for differential equations, discussed in Sect. 8.2: a difference equation is chaotic if it shows SDIC but has bounded solutions. The SDIC can be quantified by a Lyapunov exponent, which measures the rate of exponential separation. The discrete equivalent of (8.15) is

$$|x_n(x_0 + \varepsilon) - x_n(x_0)| \propto e^{\Lambda n}, \tag{8.23}$$

where the notation $x_n(x_0)$ indicates the value of x_n when the initial condition for the difference equation is x_0, and ε is very small. If $\Lambda > 0$, then nearby points separate exponentially, so as long as the system is bounded, it is chaotic. The form of (8.23) is suggestive of a derivative, so

$$\left| \varepsilon \frac{dx_n}{dx_0} \right| \propto e^{\Lambda n}.$$

Taking the logarithm and the limit $n \to \infty$ leads to the definition

$$\Lambda = \lim_{n \to \infty} \frac{1}{n} \log \left| \frac{dx_n}{dx_0} \right|.$$

Since x_n depends on x_{n-1} which in turn depends on x_{n-2}, and so on, Λ can be written as

$$\Lambda = \lim_{n \to \infty} \frac{1}{n} \log \left| \frac{dx_n}{dx_{n-1}} \frac{dx_{n-1}}{dx_{n-2}} \cdots \frac{dx_1}{dx_0} \right| = \lim_{n \to \infty} \frac{1}{n} \log \left| \prod_{j=0}^{n-1} \frac{dx_{j+1}}{dx_j} \right| \tag{8.24}$$

using the chain rule $n - 1$ times, where the $\prod$ symbol represents the product. Now using the facts that $|ab| = |a||b|$ and $\log(|a||b|) = \log|a| + \log|b|$,

$$\Lambda = \lim_{n \to \infty} \frac{1}{n} \sum_{j=0}^{n-1} \log \left| \frac{dx_{j+1}}{dx_j} \right| = \lim_{n \to \infty} \frac{1}{n} \sum_{j=0}^{n-1} \log \left| 1 + f'(x_j) \right| \tag{8.25}$$

for the difference equation (7.1). The Lyapunov exponent has a straightforward interpretation and is easy to compute: it is the long-term average of logarithm of the modulus of the derivative of the difference equation. The average is taken along a particular trajectory, and any chaotic trajectory will give the same answer. This formula (8.25) means that computing the Lyapunov exponent for a difference equation is much easier than for a system of differential equations.

A number of simple results follow from (8.25).

1. If the iterations of a difference equation approach a stable fixed point x_*, then $\Lambda < 0$, because the long-term average in (8.25) is just $\log|1 + f'(x_*)|$ which is negative, because for a stable fixed point $-2 < f'(x_*) < 0$ so the quantity inside the logarithm is between 0 and 1.
2. If the iterations approach a stable periodic cycle, then $\Lambda < 0$. This follows from the previous result because a stable cycle of period m is a stable fixed point of the difference equation applied m times.
3. If a fixed point is superstable, Λ is undefined, and $\Lambda \to -\infty$ as a parameter is varied so that the system approaches the superstable fixed point.
4. As a superstable point of a periodic cycle is approached, $\Lambda \to -\infty$.
5. At a bifurcation point of a fixed point, $\Lambda = 0$. This is because $f'(x_*) = 0$ or -2 at a bifurcation, so the quantity inside the logarithm is 1.

The Lyapunov exponent for $x_{n+1} = x_n + \mu - x_n^2$ is plotted as a function of μ in Fig. 8.12, computed by using the formula (8.25) with $n = 1000$. At the period-doubling bifurcations, for example $\mu = 1$ and $\mu = 1.5$, $\Lambda = 0$ as shown above. For

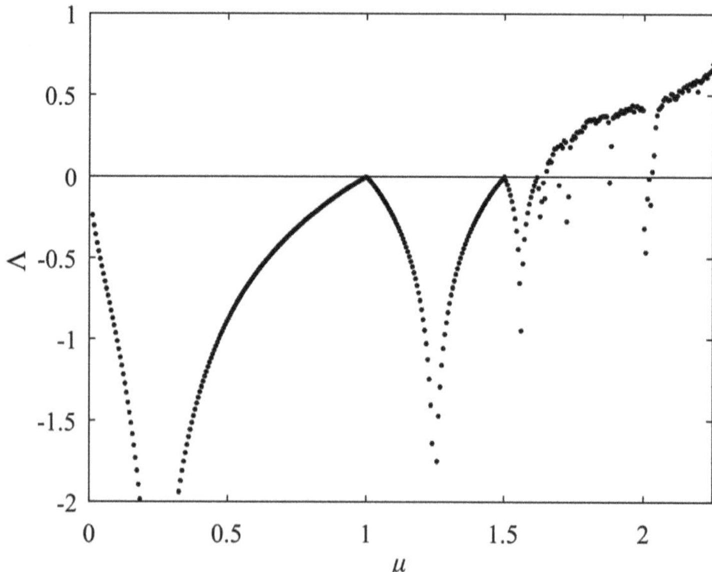

Fig. 8.12 Lyapunov exponent $\Lambda(\mu)$ for $x_{n+1} = x_n + \mu - x_n^2$

$\mu < 1$, where the fixed point $x_* = \sqrt{\mu}$ is stable, $\Lambda = \log|1-2x_*| = \log|1-2\sqrt{\mu}|$. This tends to $-\infty$ as $\mu \to 1/4$, where x_* is superstable. The other sharp downward spikes in the diagram are points where a periodic state is superstable. Chaos sets in around $\mu \approx 1.65$, and then Λ increases with μ, except for the windows of stable periodic behaviour where $\Lambda < 0$.

8.3.5   Lorenz's Tent Map

This section and the following one introduce two simple but important examples of chaos in discrete systems. These are useful because many of the results about chaos, that have been stated but not demonstrated in previous sections, can easily be proved for these systems. Both of these are more naturally expressed in the form of a *map*, in the form (7.3),

$$x_{n+1} = F(x_n), \tag{8.26}$$

rather than as a difference equation (7.1), although as discussed in Remark 7.1, any map can be written as a difference equation and vice versa.

In his famous 1963 paper, Lorenz introduced the following method of understanding the irregular oscillations of his three equations (8.7)–(8.9). Following the variable $z(t)$, he computed the maximum values of z, giving a list of values Z_n. Then he plotted Z_{n+1} as a function of Z_n. Remarkably, he found that the values lay on a curve, or rather, two symmetrical curves either side of a sharp peak. This function and the associated map is known as the *Lorenz map*. Figure 8.13 shows the Lorenz

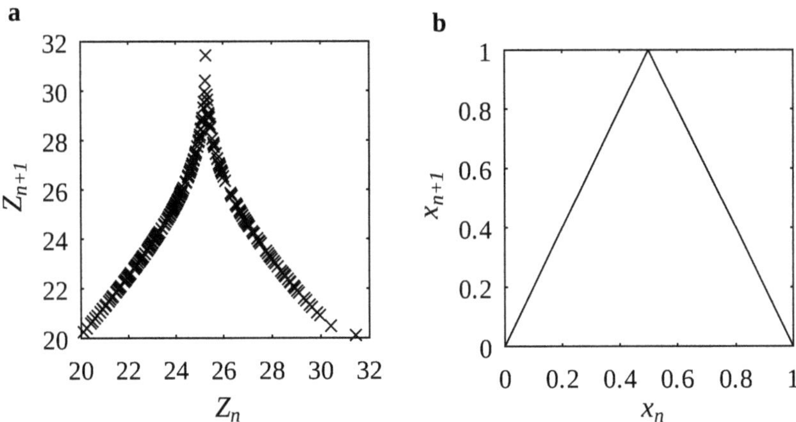

Fig. 8.13 (a) The Lorenz map, Z_{n+1} plotted as a function of Z_n, where the Z_n are the maximum values of z, for $\sigma = 5, r = 20$. (b) The tent map (8.27)

map for $\sigma = 5, r = 20$. The appearance of the graph is similar for other choices of σ and r in the chaotic regime.

To investigate this further, Lorenz introduced a simplified version of this diagram, with the curves replaced by two straight lines, which is known as the *tent map*, for obvious reasons:

$$x_{n+1} = \begin{cases} 2x_n & \text{for } 0 \le x_n < 1/2, \\ 2(1 - x_n) & \text{for } 1/2 \le x_n \le 1. \end{cases} \tag{8.27}$$

Any starting value x_0 with $0 \le x_0 \le 1$ generates an infinite sequence of values x_n that also lie in [0, 1].

The action of the tent map on the interval [0, 1] is exactly the stretching and folding mechanism shown in Fig. 8.8. The interval [0, 1/2) is stretched out by a factor 2 to cover the interval [0, 1), and the interval [1/2, 1] is stretched out by a factor of 2 but reversed, or folded back, so that 1/2 maps to 1 and 1 maps to zero.

Since $|dx_{n+1}/dx_n| = 2$ always, it follows directly from (8.25) that the Lyapunov exponent for the tent map is $\Lambda = \log 2$. This is positive, so the tent map is chaotic.

The tent map has two fixed points, where $x_{n+1} = x_n$. These are $x = 0$ and $x = 2/3$. Both of these are unstable.

Now consider the behaviour of (8.27) for different choices of x_0. Suppose first that $x_0 = p/2^n$, where p and n are positive integers, and we may assume that p is odd. Then x_1 is either $p/2^{n-1}$ or $2 - p/2^{n-1}$, and in each case the denominator is 2^{n-1}. After each iteration, the power of 2 in the denominator is decreased by one, until after n steps we have $x_n = 1$, and then $x_k = 0$ for all $k > n$. So all initial conditions of this type approach the (unstable) fixed point $x = 0$. Similarly, there are infinitely many initial values where the iterations end up at the other fixed point, 2/3.

For the initial condition $x_0 = 2/5$, $x_1 = 2 \times 2/5 = 4/5$ and $x_2 = 2 \times (1 - 4/5) = 2/5 = x_0$, so we have a period-two cycle. In fact this is the only period-two cycle. More generally, suppose that $x_0 = p/q$ where p and q are positive integers with $p < q$, and q is odd. Then $x_1 = 2p/q$ or $2 - 2p/q$, and in either case the denominator is still q. Since there are only finitely many possible fractions p/q in the interval [0, 1], this initial condition must either approach the fixed point 2/3, which is only possible if $q = 3$, or a periodic cycle, for all odd $q > 3$.

If q is an even number, then either the sequence of iterations leads to 0, if q is a power of 2, as already discussed, or a periodic state is found involving fractions with an odd denominator. For example, if $x_0 = 7/20$, then the sequence is $7/20 \rightarrow 7/10 \rightarrow 3/5 \rightarrow 4/5 \rightarrow 2/5 \rightarrow 4/5$, so this initial condition leads to the period-two cycle. Therefore, every rational initial condition leads to either a periodic orbit or the fixed point $x = 0$.

For any real number, there is a rational number arbitrarily close to it. Therefore, there exists a periodic orbit of the tent map arbitrarily close to any point in the interval [0, 1]. In other words, since rational numbers are dense in the real numbers, periodic orbits are dense in the tent map. So we have demonstrated that the tent map satisfies one of the properties of chaotic systems discussed in Sect. 8.2.

Example 8.4 Find all the period-three cycles of the tent map.

There are two possible types of period-three orbit, depending on which half of the interval each point lies in. These can be described using a symbolic notation like that used to describe periodic orbits of the Lorenz equations. For example, LLR describes a cycle where x_0 and x_1 are less than 1/2 but $x_2 > 1/2$. For a cycle of this type, $x_1 = 2x_0$, $x_2 = 2x_1 = 4x_0$ and $x_3 = x_0 = 2(1 - x_2) = 2(1 - 4x_0)$. Solving this equation for x_0 gives $x_0 = 2/9$, so $x_1 = 4/9$ and $x_2 = 8/9$. The other possible cycle type is LRR, which leads to $x_1 = 2x_0$, $x_2 = 2(1 - x_1) = 2 - 4x_0$ and $x_0 = 2(1 - (2 - 4x_0)) = -2 + 8x_0$, so $x_0 = 2/7$, $x_1 = 4/7$, $x_2 = 6/7$.

Remark 8.2 The choice of a period-three cycle in Example 8.4 is significant, because the explicit construction of period-three cycles shows, thanks to Sharkovsky's theorem, that periodic orbits of all other periods exist in the tent map. These can be constructed explicitly, as shown in the following example.

Example 8.5 Construct a period-n cycle of the tent map.

We have already seen the two-cycle $(2/5, 4/5)$, of type LR and the three-cycle $(2/9, 4/9, 8/9)$, of type LLR. This suggests seeking an n-cycle of the form $(2/q, 4/q, 8/q \ldots 2^n/q)$. This is a solution of the tent map of the form LLLL...LLLR if $2^n/q < 1$ and $2(1 - 2^n/q) = 2/q$. Solving for q in terms of n gives $q = 2^n + 1$. Hence there is a 4-cycle with $x_0 = 2/17$, a 5-cycle with $x_0 = 2/33$, and so on. There are many other periodic cycles of different symbolic type.

8.3.6 The Doubling Map

There is one chaotic discrete dynamical system that is even simpler than the tent map, where the behaviour is easier to understand. Consider

$$x_{n+1} = 2x_n \bmod 1 = \begin{cases} 2x_n & \text{for } 0 \le x_n < 1/2, \\ 2x_n - 1 & \text{for } 1/2 \le x_n < 1. \end{cases} \tag{8.28}$$

We will call this the doubling map, though it is also known as the shift map or $2x$ mod 1 map. Clearly this is closely related to the tent map (8.27). However, whereas the tent map models the stretching and folding process shown in Fig. 8.8, the doubling map describes a stretch, cut, slide process: the interval $[0, 1)$ stretches out by a factor of two to $[0, 2)$, splits into two, and then the upper half slides back over the lower half.

Since the derivative of the map is 2 at all points where it is defined, the Lyapunov exponent is $\log 2$, as for the tent map, and the dynamics is bounded, so (8.28) is chaotic. The only fixed point is $x = 0$, which is unstable, since for any small x_0,

Table 8.1 Action of the doubling map (8.28) on binary numbers

	$x_0 = 37/64$	$x_0 = 37/63$
x_0	0.100101	$0.\overline{100101}$
x_1	0.001010	$0.\overline{001011}$
x_2	0.010100	$0.\overline{010110}$
x_3	0.101000	$0.\overline{101100}$
x_4	0.010000	$0.\overline{011001}$
x_5	0.100000	$0.\overline{110010}$
x_6	0.000000	$0.\overline{100101}$

$x_1 = 2x_0$, $x_2 = 4x_0$, so the iterations move away from 0. Any periodic orbit must also be unstable for the same reason: any small perturbation is doubled at each step.

The behaviour of (8.28) is best understood by writing the initial condition x_0 in binary form. Multiplication of a binary number by 2 just moves the bits (binary digits) one place to the left, and the mod 1 operation means that if there is a 1 to the left of the decimal point, this is discarded. This is why (8.28) is sometimes called the shift map. Table 8.1 shows the iteration sequence from the initial condition $x_0 = 0.100101$ and for $x_0 = 0.\overline{100101}$, where the overbar notation indicates that the bits are repeated. Using this binary notation, a number of results about the behaviour of the map can easily be deduced.

1. Any initial condition with a finite binary representation leads to the unstable fixed point at 0. These are initial conditions of the form p/q where q is a power of 2, as for the tent map.
2. There is sensitive dependence on initial conditions, because any small change in x_0 is repeatedly doubled. If the $n + 1$th bit after the decimal point in x_0 is 0, then x_n lies in the interval $[0, 1/2]$, but if this bit is 1 then x_n is in the interval $[1/2, 1]$.
3. There are orbits of any period n, corresponding to binary numbers x_0 that repeat after n bits. These are rational numbers, since they have the property that $2^n x_0 = m + x_0$, where m is the integer corresponding to the binary number represented by the repeating bits. For example, $x_0 = 0.\overline{1001}$ lies on a period-four orbit, and $16x_0 = 9 + x_0$, so $x_0 = 3/5$ and the orbit is $(3/5, 1/5, 2/5, 4/5)$. For $x_0 = 0.\overline{100101}$, the orbit period is 6 and $64x_0 = 37 + x_0$, so $x_0 = 37/63$, see Table 8.1.
4. If p is a prime number, then the number of orbits of period p is $(2^p - 2)/p$. This is because there are 2^p possible arrangements of 0 and 1 of length p. This includes $0.\overline{0}$ and $0.\overline{1}$ which do not have period p, so these two must be subtracted off. An orbit of period p contains p of these arrangements, so to find the number of distinct orbits we need to divide $2^p - 2$ by p. So there are $(128 - 2)/7 = 18$ different orbits of period 7, and 630 orbits of period 13. If p is not a prime, the calculation is more complicated, as we have to subtract off the orbits with length equal to a factor of p.
5. Any initial condition that eventually repeats itself leads to a periodic orbit. For example $x_0 = 0.101100\overline{101}$ joins a period-three orbit after 6 steps.

6. Periodic orbits are dense: for any x and any $\varepsilon > 0$, there is a point x_0 on a periodic orbit with $|x - x_0| < \varepsilon$. This can be proved by construction. Given x and ε, choose $m > -\log \varepsilon / \log 2$, so that $2^{-m} < \varepsilon$, and let x_0 be the repeating binary number using the first m bits of x. Then x_0 lies on an orbit of period m and $|x - x_0| < 2^{-m} < \varepsilon$.

7. Periodic orbits can be described using the same symbolic notation as for the Lorenz equations and the tent map, and the symbolic form of an orbit can be read off from the binary notation. For example, $0.\overline{100101}$ is on an orbit of type RLLRLR.

Key Points from Chap. 8

- The phenomenon of *chaos* can only occur in differential equations of order three or higher.
- The distinctive feature of chaos is *sensitive dependence on initial conditions*: adjacent trajectories diverge from each other exponentially fast. However, all solutions remain in a finite region of phase space. A useful way to think of chaos is as a repeated stretching and folding of phase space.
- The trajectories of a chaotic system form a complicated structure known as a *strange attractor*, that is difficult to visualise.
- The Lorenz equations are one of the simplest (yet in fact extremely complicated) systems of three coupled differential equations illustrating chaos. They were designed as a model of convection currents in the atmosphere.
- The sensitive dependence on initial conditions means that the behaviour of chaotic systems is very hard to predict. This is one of the reasons why accurate weather forecasting is so difficult.
- Chaos is very difficult to analyse in differential equations.
- Chaos is much easier to study in discrete systems. It can occur in the simple first-order iteration $x_{n+1} = f(x_n)$. Two simple examples are the tent map and the doubling map.
- Chaos can arise via an infinite sequence of period-doubling bifurcations as a parameter varies. The sequence converges to a limit in a way that is *universal*—essentially the same in all systems. If the parameter is increased beyond this limit, chaos occurs.
- Chaos can be quantified by the Lyapunov exponent, which measures the exponential separation rate of two nearby solutions.
- If a discrete dynamical system has a cycle of period three, it also has cycles of any order.
- In a chaotic system, periodic orbits are dense: any point is arbitrarily close to a point on a periodic orbit.

Exercises

8.1 Evaluate the integrals

$$S_m = \int_0^{2\pi} \sin\theta \sin(m\theta)\, d\theta, \qquad C_m = \int_0^{2\pi} \sin\theta \cos(m\theta)\, d\theta,$$

where m is an integer, in the cases $m = 1$ and $m \neq 1$. Hence justify the derivation of (8.4) from (8.1).

8.2 Derive the Lorenz equations for the waterwheel model shown in Fig. 8.1b, assuming that the distribution of water around the wheel can be described by a continuous function $m(\theta, t)$, so that the mass of water between θ and $\theta + \delta\theta$ is $m(\theta, t)\delta\theta$.

8.3 Following the method of Sect. 8.1.2, show that solutions of the Lorenz equations are bounded by considering the function

$$V(x, y, z) = \frac{r}{\sigma}x^2 + y^2 + (z - 2r)^2.$$

8.4 (a) Derive the characteristic equation (8.12) for the stability of the two non-zero fixed points of the Lorenz equations.

(b) Confirm that for $r = 1 + \varepsilon$, $0 < \varepsilon \ll 1$, these fixed points are stable. Hint: for small ε, two of the eigenvalues are $O(1)$ and the other is $O(\varepsilon)$.

8.5 What value of σ in the Lorenz equations minimises the value of r_H, where the Hopf bifurcation occurs? What is the minimum value of r_H?

8.6 Find Lorenz's 1963 paper and read it. It's very well written and not difficult to read. Can you find a typo?

8.7 Write your own computer code to simulate the Lorenz equations, either using Euler's method as in the pseudocode in Sect. 8.1.4, or using a more sophisticated time-stepping method.

(a) Run your code for $\sigma = 5$, $r = 5$, and check that it gives the correct behaviour. What happens as the step size h is increased?

(b) Check your code for $r = r_H - 1$ and for $r = r_H + 1$, with an initial condition close to one of the non-zero fixed points. You may need to reduce h to get the correct behaviour.

(c) Investigate how solutions depend on the initial condition, for $r = 14$ and $r = 13$. Use the results from (b) to guide your choice of step size.

(d) Figure 8.2 was generated using the initial condition $x = 1$, $y = -1$, $z = 11$, with $r = 17$. Can you reproduce it with your code? If not, why not?

8.8 Using your Lorenz equations code, investigate solutions for $\sigma = 5$, $70 < r < 80$. For these large values of r, if using Euler's method, you need a very small step size, around 10^{-4}, to get accurate solutions. Can you find a window of stable periodic solutions? If so, is the periodic orbit symmetrical or not, and what is the type of the orbit, in the notation used at the end of Sect. 8.1.4?

8.9 Following the method of Example 8.1, find the eigenvalues and eigenvectors of the non-zero fixed points of the Lorenz equations at the Hopf bifurcation for a general value of σ. Hence find the equation of the plane in which the Hopf bifurcation occurs. Check that your general result is consistent with Example 8.1.

8.10 Investigate the Rössler equations, given in Sect. 8.2, as follows.

(a) Consider the existence of fixed points, and hence deduce that there are two stationary bifurcations and determine their type.
(b) Modify your code for the Lorenz equations to solve the Rössler equations. Fix $b = 3$ and investigate increasing a from -0.5 to 0.5. Describe the bifurcations and types of solutions that you find.
(c) Derive the first bifurcation that you found in (b) analytically.

8.11 Compute the bifurcation diagrams for the difference equations

(a) $x_{n+1} - x_n = e^{-x_n} - \mu$
(b) $x_{n+1} - x_n = \mu - x_n^4$
(c) $x_{n+1} - x_n = -\mu \log(x_n)$
(d) $x_{n+1} - x_n = 1 + \mu \sin(x_n)$.

Which of these have bifurcation diagrams qualitatively similar to Fig. 8.9?

8.12 For the difference equation (7.19), the first period-doubling bifurcation takes place at $\mu = \mu_1 = 1$ and the second at $\mu_2 = 1.5$. Use your difference equation code to find the location of the next three period-doubling bifurcations, μ_3, μ_4 and μ_5. You will need to increase the number of iterations used to find the higher bifurcations accurately. Find the ratio of gaps between successive bifurcations and compare your result with the Feigenbaum constant.

8.13 Figure 8.9 suggests that in addition to the period-three window near $\mu = 2$, the difference equation $x_{n+1} - x_n = \mu - x_n^2$ may have other windows of periodic behaviour. Use your difference equation code to look for such windows in the range $1.6 < \mu < 2$. For each window you find, give the length of the periodic cycle and a value of μ at which the cycle is stable.

8.14 Find the Sharkovsky ordering of the integers $60, 61 \ldots 70$.

8.15 (a) Using the method of Example 8.3, find the region of existence of a period-three solution of $x_{n+1} - x_n = -\mu x_n/(1 + x_n)$.
(b) At the saddle–node bifurcation of period-three solutions, show that one of the points in the cycle is $x_0 = 1 + 2\cos(\pi/9)$ and find the other two points.

8.16 Find the formula for $\Lambda(\mu)$ for $1 < \mu < 1.5$ in Fig. 8.12. For what value of μ is the period-two cycle superstable?

8.17 Write a computer code to run the tent map (8.27) for 100 iterations. Starting from a random initial condition, what happens? Also try the initial condition $x_0 = 1/7$. Can you explain the results?

8.18 How many distinct period-four cycles does the tent map have? Write them down in the symbolic notation used in Example 8.4 and write down a fraction that generates each of them.

8.19 Investigate computationally the behaviour of the tent map for the initial conditions $x_0 = 0.23$ and $x_0 = 0.51$, and other initial conditions containing only two decimal places. What do you find, and can you explain your results?

8.20 Consider the parameterised form of the tent map,

$$x_{n+1} = \begin{cases} \mu x_n & \text{for } 0 \le x_n < 1/2, \\ \mu(1 - x_n) & \text{for } 1/2 \le x \le 1. \end{cases}$$

(a) For what values of $\mu > 0$ is this map bounded for initial conditions in $[0, 1]$?
(b) For what values is it chaotic?
(c) Restricting attention to where solutions are bounded and chaotic, find the values of μ for which a period-three solution of type LLR exists. What can be deduced from this result?

8.21 Suppose that the fraction p/q lies on a period-five orbit of the doubling map. What is the value of q? Hence deduce, in binary form, representatives of each of the possible period-five orbits. Check that the total number is consistent with the formula given in point 4 of Sect. 8.3.6.

8.22 Consider the map

$$x_{n+1} = 3x_n \text{ mod } 1. \tag{8.29}$$

(a) Find the fixed points.
(b) Find a period-three orbit.
(c) How many period-three orbits are there? How many period-seven orbits?
(d) Modify your computer code from Exercise 8.17 to investigate (8.29), carrying out 100 iterations. Run your code with a random initial condition, and also use it to check your answer to part (b). Discuss the results, and compare them with your results from Exercise 8.17.

8.23 A parameterised form of the doubling map is

$$x_{n+1} = \mu x_n \text{ mod } 1, \quad \text{where } 1 \le \mu \le 2. \tag{8.30}$$

(a) Find the period-two solution, and find the range of μ values for which it exists.
(b) Find the values of μ for which a period-three cycle of type LLR exists. Hint: the only root of the cubic equation $z^3 - z^2 - 1 = 0$ is $z \approx 1.4656$.
(c) Explain why the results of (a) and (b) do not contradict Sharkovsky's theorem.

Chapter 9
Solutions to Odd-Numbered Exercises

Problems of Chap. 2

2.1

(a) Second-order, linear, inhomogeneous.
(b) First-order, nonlinear.
(c) Third-order, nonlinear.
(d) Second-order, linear, homogeneous.

2.3 This equation is separable, so it is solved by

$$\int \frac{x}{x^2+1}\, dx = \int t\, dt \;\Rightarrow\; \frac{1}{2}\log(x^2+1) = \frac{1}{2}t^2 + C$$

$$\Rightarrow\; x^2 + 1 = Ae^{t^2} \;\Rightarrow\; x = \sqrt{Ae^{t^2} - 1},$$

where $A = e^{2C}$. From the initial condition $x(0) = 1$ it follows that $A = 2$. If the initial condition is $x(0) = -1$, everything is the same, including the value of A, but we must be careful when taking the square root and choose the negative square root to obey the condition, so $x = -\sqrt{2e^{t^2} - 1}$.

2.5 For (2.19), the auxiliary equations has two equal roots if $b^2 = 4ac$, and the double root is $m = -b/(2a)$, with $am^2 + bm + c = 0$. Setting $y = Ae^{mx} + Bxe^{mx}$ in (2.19) gives $A(am^2 + bm + c)e^{mx} + B(a(2m + m^2x) + b(1 + mx) + cx)e^{mx} = B(2am + b)e^{mx} = 0$, so (2.19) is satisfied.

Supplementary Information The online version contains supplementary material available at (https://doi.org/10.1007/978-3-031-99543-9_9).

    197
P. C. Matthews, *Differential Equations, Bifurcations and Chaos*,
Springer Undergraduate Mathematics Series,
https://doi.org/10.1007/978-3-031-99543-9_9

2.7 Following the same procedure as for a second-order equation, set $y = Ae^{mx}$, leading to $m^4 - 13m^2 + 36 = 0$. This quartic for m can be solved because it is just a quadratic for m^2, and it factorises into $(m^2 - 9)(m^2 - 4) = 0$. So the four solutions for m are $m = \pm 3, \pm 2$ and the general solution is

$$y = Ae^{3x} + Be^{-3x} + Ce^{2x} + De^{-2x},$$

for arbitrary constants A, B, C, D.

2.9 To eliminate x, differentiate (2.26) and substitute to remove $\dot{x}$ and then again to remove x. This gives

$$\ddot{y} - 5\dot{y} + 6y = 0.$$

Note that this is the same equation as was found for the variable x when y was eliminated. Hence the general solution is $y = Ce^{2t} + De^{3t}$, and $x = \dot{y} - 4y = -2Ce^{2t} - De^{3t}$. For the initial conditions $x(0) = 2$, $y(0) = 5$, $-2C - D = 2$ and $C + D = 5$, so $C = -7$ and $D = 12$, and the specific solution is $x = 14e^{2t} - 12e^{3t}$, $y = -7e^{2t} + 12e^{3t}$.

2.11 To solve $y' + p(x)y = q(x)y^n$, set $u = y^{1-n}$, then

$$u' = (1-n)y^{-n}y' = (1-n)y^{-n}(-p(x)y + q(x)y^n) = (n-1)p(x)u - (n-1)q(x).$$

This is a first-order linear equation for $u(x)$.

For $xy' + y = 2x^3/y^2$, $n = -2$ so the substitution is $u = y^3$ and the resulting linear equation is $u' + (3/x)u = 6x^2$. Multiplying by the integrating factor x^3 gives $x^3 u' + 3x^2 u = 6x^5$ so the solution for u is $u = x^3 + C/x^3$, where C is any constant. Hence the solution for $y(x)$ is $y = \left(x^3 + C/x^3\right)^{1/3}$.

Problems of Chap. 3

3.1 $\dot{x} = x^2 - x - 6 = (x-3)(x+2)$ so there are fixed points at $x = 3$ and $x = -2$. If $f(x) = x^2 - x - 6$ then $f'(x) = 2x - 1$. $f'(3) = 5$ so $x = 3$ is unstable, and $f'(-2) = -5$ so $x = -2$ is stable.

3.3

(a) For $\dot{x} = ax - bx^3$, rescale $x = \alpha X$, $t = \beta T$, as in Example 3.4. Then $(\alpha/\beta)\dot{X} = a\alpha X - b\alpha^3 X^3$. There are no constants on the RHS if we choose $\alpha^2 = a/b$, which we can do, as we are told that $a > 0$ and $b > 0$. To remove the constants on the LHS we need $\beta = 1/a$. Then we have a dimensionless equation $\dot{X} = X - X^3$.

(b) The fixed points are at $X = 0$ and $X = \pm 1$. For $f(X) = X - X^3$, $f'(x) = 1 - 3X^2$. Since $f'(0) = 1 > 0$, the point $X = 0$ is unstable. But $f'(\pm 1) = -2$ so these points are stable. For a population model, only positive solutions are of interest. All solutions with $X(0) > 0$ tend to the stable fixed point $X = 1$.

If $X(0) < 1$ then X increases and approaches 1 from below as $t \to \infty$. If $X(0) > 1$, X decreases and tends to 1.

(c) $\dot{X} = X - X^3$ is separable. Use partial fractions to do the integral:

$$\int \frac{dX}{X - X^3} = \int dt = \int \frac{A}{X} + \frac{B}{1 - X} + \frac{C}{1 + X} \, dx.$$

To find A, B and C, multiply up to get $1 = A(1-X)(1+X)+B(1+X)X+C(1-X)X$, for all X. Choose $X = 0$ to find $A = 1$, then $X = 1$ to get $B = 1/2$ and $X = -1$ to show $C = -1/2$. Hence $\log|X| - \log|1 - X|/2 - \log|1 + X|/2 = t + c$, so $\log|X/\sqrt{(1 + X)/(1 - X)}| = t + c$ and $X^2/(1 - X^2) = e^{2t+2c}$ which after rearranging gives $X = 1/\sqrt{1 + De^{-2t}}$. Hence $X \to 1$ as $t \to \infty$. If the constant D is positive then X tends to 1 from below. If $-1 < D < 0$, X approaches 1 from above. This is all consistent with the qualitative approach used in (b). The point of the question is that finding the exact solution is much more work but not much more informative than the qualitative solution.

3.5

(a) For (3.13), $\dot{y} = y^3 f'''(x_0)/6$, with initial condition $y(0) = y_0$, the equation is separable and using the method of Sect. 2.2 gives the solution

$$y = \frac{y_0}{\sqrt{1 - ty_0^2 f'''(x_0)/3}}.$$

(b) In the stable case, $f'''(x_0) < 0$ so y is proportional to $t^{-1/2}$ as $t \to \infty$.

(c) In the unstable case, $f'''(x_0) > 0$, the solution tends to infinity as t tends to a finite time, $3/(y_0^2 f'''(x_0))$.

3.7 For $\dot{x} = f(x)$, replacing x by $-x$ gives $-\dot{x} = f(-x)$, so $\dot{x} = -f(-x)$. If this is the same as the original equation, then $f(x) = -f(-x)$, so f is an odd function. Setting $x = 0$ here gives $f(0) = -f(0)$, so $f(0) = 0$ and hence $x = 0$ is a fixed point. If the symmetry is $x \leftrightarrow c - x$ then $-\dot{x} = f(c - x)$ so $f(x) = -f(c - x)$. Now put $x = c/2$ to see that $f(c/2) = -f(c/2)$ so $f(c/2) = 0$, and there is a fixed point at $x = c/2$.

3.9 $\dot{x} = t - x$ can be written $\dot{x} + x = t$. This is a first-order linear equation. The integrating factor is e^t and multiplying through by this factor gives

$$\frac{d(xe^t)}{dt} = te^t \implies xe^t = \int te^t \, dt = te^t - e^t + c,$$

using integration by parts, where c is any constant. Hence the general solution is $x = t - 1 + ce^{-t}$, so for large t, x approaches the line $x = t - 1$.

Problems of Chap. 4

4.1

(a) Define $\dot{x} = y$, then the original equation is $\dot{y} = 2y - 2x$, so we have two first-order equations for x and y.

(b) Set $y = x - \dot{x}$. Then $\dot{y} = \dot{x} - \ddot{x} = \dot{x} - (2\dot{x} - 2x) = -\dot{x} + 2x = x + (x - \dot{x}) = x + y$. So the coupled system is $\dot{x} = x - y$, $\dot{y} = x + y$. Note that there are many different ways of splitting a second-order equation into two first-order ones.

(c) For parts (a) and (b) the matrices are M_a and M_b, where

$$M_a = \begin{pmatrix} 0 & 1 \\ -2 & 2 \end{pmatrix}, \qquad M_b = \begin{pmatrix} 1 & -1 \\ 1 & 1 \end{pmatrix}.$$

Both matrices have trace $T = 2$ and determinant $D = 2$ so they have the same eigenvalues, the roots of the quadratic $\lambda^2 - 2\lambda + 2 = 0$, so $\lambda = 1 \pm i$. This is to be expected—the exponential solutions of the original equation should not depend on the choice of how it is split into two equations.

(d) The eigenvalues are complex with $Re(\lambda) > 0$, so it is an unstable spiral.

4.3

(a) This system is uncoupled, but can still be written in matrix form as

$$\begin{pmatrix} \dot{x} \\ \dot{y} \end{pmatrix} = \begin{pmatrix} 1 & 0 \\ 0 & -1 \end{pmatrix} \begin{pmatrix} x \\ y \end{pmatrix}.$$

The eigenvalues and eigenvectors are $\lambda = 1$, $(1, 0)^T$ and $\lambda = -1$, $(0, 1)^T$. The system is a saddle.

(b) In matrix form,

$$\begin{pmatrix} \dot{x} \\ \dot{y} \end{pmatrix} = \begin{pmatrix} 0 & 2 \\ -3 & 0 \end{pmatrix} \begin{pmatrix} x \\ y \end{pmatrix}.$$

The matrix has $T = 0$ and $D = 6$, so it represents a centre, with eigenvalues $\lambda = \pm\sqrt{6}i$. When $y > 0$, $\dot{x} > 0$, so the rotation is clockwise.

(c)

$$\begin{pmatrix} \dot{x} \\ \dot{y} \end{pmatrix} = \begin{pmatrix} -3 & 1 \\ -2 & 0 \end{pmatrix} \begin{pmatrix} x \\ y \end{pmatrix} \Rightarrow T = -3, D = 2.$$

$T^2 - 4D = 1 > 0$ so we have a stable node. Eigenvalues obey $\lambda^2 + 3\lambda + 2 = 0$ so $\lambda = -1$ or $\lambda = -2$. Eigenvectors are any multiple of

$$\begin{pmatrix} 1 \\ 2 \end{pmatrix} \text{ for } \lambda = -1, \qquad \begin{pmatrix} 1 \\ 1 \end{pmatrix} \text{ for } \lambda = -2.$$

(d)

$$\begin{pmatrix} \dot{x} \\ \dot{y} \end{pmatrix} = \begin{pmatrix} 5 & 4 \\ -1 & 1 \end{pmatrix} \begin{pmatrix} x \\ y \end{pmatrix} \quad \Rightarrow \quad T = 6, \ D = 9.$$

Characteristic equation is $\lambda^2 - 6\lambda + 9 = 0 \Rightarrow (\lambda - 3)^2 = 0$. There is only one eigenvalue $\lambda = 3$ so this system is a degenerate node. Since $\lambda > 0$ it is an unstable degenerate node. The eigenvector is $(2, -1)^T$.

(e)

$$\begin{pmatrix} \dot{x} \\ \dot{y} \end{pmatrix} = \begin{pmatrix} 4 & -9 \\ 1 & 8 \end{pmatrix} \begin{pmatrix} x \\ y \end{pmatrix} \quad \Rightarrow \quad T = 12, \ D = 41.$$

$T^2 - 4D = -20 < 0$ and $T > 0$ so this is an unstable spiral. Since $\dot{x} < 0$ at $(0, 1)$, and $\dot{y} > 0$ at $(1, 0)$, the trajectories rotate anticlockwise.

4.5 The system has trace $T = 4 - 7 = -3$ and determinant $D = -28 + 30 = 2$ so the characteristic equation is $\lambda^2 + 3\lambda + 2 = 0$ and the eigenvalues are $\lambda = -2$ and $\lambda = -1$. The system is a stable node, so all solutions approach the origin. The eigenvectors are

$$\begin{pmatrix} 1 \\ 2 \end{pmatrix} \text{ for } \lambda = -2 \quad \text{and} \quad \begin{pmatrix} 3 \\ 5 \end{pmatrix} \text{ for } \lambda = -1.$$

The general solution with two arbitrary constants is

$$\begin{pmatrix} x \\ y \end{pmatrix} = A e^{-2t} \begin{pmatrix} 1 \\ 2 \end{pmatrix} + B e^{-t} \begin{pmatrix} 3 \\ 5 \end{pmatrix}.$$

Since e^{-2t} decays more rapidly than e^{-t}, solution trajectories become increasingly aligned with the eigenvector corresponding to $\lambda = -1$ as they approach the origin, that is, along the line $y = 5x/3$.

This will occur unless $B = 0$. In this special case, solutions approach the origin along the straight line $y = 2x$. This corresponds to initial conditions x_0, y_0 being related by $y_0 = 2x_0$.

4.7

(a) In matrix form, the system is

$$\begin{pmatrix} \dot{x} \\ \dot{y} \end{pmatrix} = \begin{pmatrix} 1 & 1 \\ 0 & 1 \end{pmatrix} \begin{pmatrix} x \\ y \end{pmatrix},$$

with eigenvalue 1 and eigenvector $(1, 0)^T$. There is only one eigenvalue and one eigenvector so it is a degenerate node.

(b) The equation for y, $\dot{y} = y$, has solution $y = y_0 e^t$, so the x equation becomes $\dot{x} = x + y_0 e^t$, which can be solved using the integrating factor e^{-t} to give $x = (y_0 t + x_0)e^t$.

(c) On any trajectory,

$$\frac{dy}{dx} = \frac{\dot{y}}{\dot{x}} = \frac{y}{x+y} = \frac{y_0}{x_0 + y_0 + y_0 t}.$$

Hence the slope of any trajectory tends to zero, aligning with the eigenvector, as $t \to \infty$ or as $t \to -\infty$.

4.9 For the matrix given in Example 4.9,

$$M = \begin{pmatrix} -1 & 1 \\ 0 & -2 \end{pmatrix} \Rightarrow M^2 = \begin{pmatrix} 1 & -3 \\ 0 & 4 \end{pmatrix} \Rightarrow M^3 = \begin{pmatrix} -1 & 7 \\ 0 & -8 \end{pmatrix},$$

so it appears that the $(1, 1)$ entry of M^n is $(-1)^n$, the $(2, 2)$ entry is $(-2)^n$, the $(1, 2)$ entry is the difference between these two, and the $(2, 1)$ entry is zero. Hence our guess is that

$$M^n = \begin{pmatrix} (-1)^n & (-1)^n - (-2)^n \\ 0 & (-2)^n \end{pmatrix}.$$

To prove that this is correct for all n using induction, first note that it is correct when $n = 1$ (the so-called "anchor step" of proof by induction). Now suppose that the formula is correct for some particular value of n, $n = N$. Then

$$M^{N+1} = \begin{pmatrix} -1 & 1 \\ 0 & -2 \end{pmatrix} \begin{pmatrix} (-1)^N & (-1)^N - (-2)^N \\ 0 & (-2)^N \end{pmatrix}$$

$$= \begin{pmatrix} -(-1)^N & -(-1)^N + (-2)^N + (-2)^N \\ 0 & -2(-2)^N \end{pmatrix}$$

$$= \begin{pmatrix} (-1)^{N+1} & (-1)^{N+1} - (-2)^{N+1} \\ 0 & (-2)^{N+1} \end{pmatrix}.$$

Therefore our guessed formula is also correct for $n = N + 1$. Hence it is true for all $n \geq 1$, by mathematical induction.

Having found M^n, the formula for the matrix exponential (4.14) can be used directly. The $(1, 1)$ entry of e^{Mt} is $1 - t + t^2/2 - t^3/6 \ldots = e^{-t}$, the $(1, 2)$ entry is $e^{-t} - e^{-2t}$, the $(2, 1)$ entry is zero and the $(2, 2)$ entry is e^{-2t}, in agreement with the result given in Example 4.9.

Fig. 9.1 Nullclines for Exercise 5.1. The dashed lines are the x-nullclines and the dash-dot lines are the y-nullclines

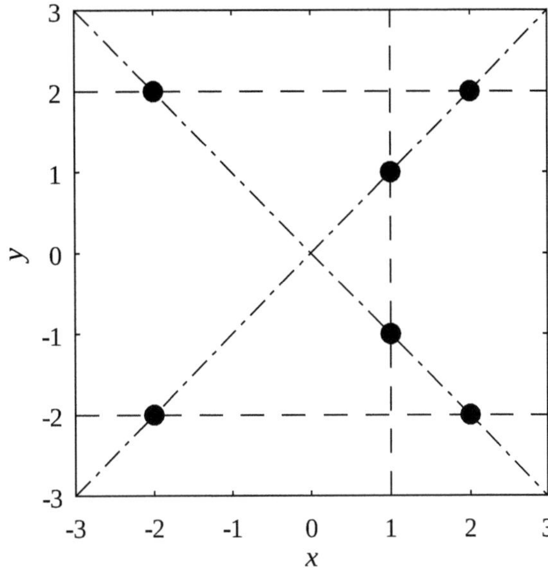

Problems of Chap. 5

5.1 The x-nullcline is where $\dot{x} = 0$, so $(x - 1)(y^2 - 4) = 0$. Hence the x-nullcline consists of three straight lines, $x = 1$, $y = 2$, $y = -2$, shown as dashed lines in Fig. 9.1. For the y-nullcline, $x^2 - y^2 = 0$, so there are two diagonal lines, $y = x$ and $y = -x$, shown as dash-dot lines. The two y-nullclines cross each of the three x-nullclines, so there are six fixed points, at $(1, \pm 1)$, $(2, \pm 2)$, $(-2, \pm 2)$, marked as large dots on the diagram. The points $(1, \pm 2)$ are *not* fixed points since the intersections there involve only two parts of the x-nullcline.

If the x equation is $\dot{x} = (x - 1)(y^2 - 1)$, the x-nullcline lines are now $x = 1$ and $y = \pm 1$, and in this case there are only four fixed points, $(\pm 1, \pm 1)$.

5.3

(a) The system $\dot{x} = y$, $\dot{y} = x(x^2 - 1)$ is Hamiltonian because the divergence condition (5.22) is satisfied. To find H, solve the equations (5.20). First $\partial H/\partial y = y$, giving $H = y^2/2 + f(x)$, and then $-\partial H/\partial x = x^3 - x$, giving $-f'(x) = x^3 - x$, so $f(x) = x^2/2 - x^4/4 + c$. We can choose $c = 0$. Hence $H(x, y) = x^2/2 + y^2/2 - x^4/4$. H is constant because $\dot{H} = (x - x^3)\dot{x} + y\dot{y} = (x - x^3)y + y(x^3 - x) = 0$.

(b) At a fixed point, $y = 0$ and $x(x^2 - 1) = 0$, so $x = 0$, $x = -1$ or $x = 1$. At $(0, 0)$, $H = 0$ and $(\pm 1, 0)$, $H = 1/4$.

(c) Hence the curve $H = 1/4$ passes through two fixed points. This curve is $x^2/2 + y^2/2 - x^4/4 - 1/4 = 0$ which can be written as $2y^2 = x^4 - 2x^2 + 1 = (x^2 - 1)^2$. Taking the square root gives two parabolas, $\sqrt{2}y = \pm(x^2 - 1)$,

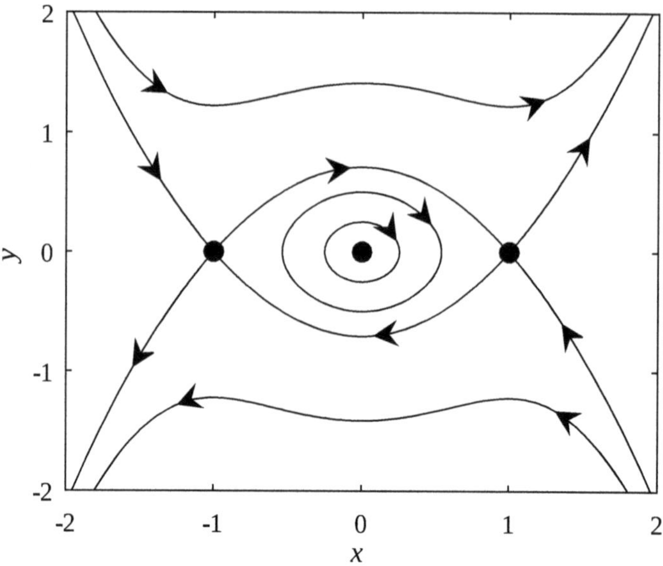

Fig. 9.2 Phase portrait for Exercise 5.3. The two parabolas passing through the fixed points $(\pm 1, 0)$ are the contours $H = 1/4$

shown in Fig. 9.2. Since $\dot{x} = y$, the direction of trajectories is always to the right when $y > 0$ and to the left when $y < 0$.

(d) Hence the two fixed points on these lines must be saddles, as shown in the figure. The remainder of the phase plane can be drawn using the fact that the trajectories are the contours $H = $ constant.

5.5 Following the hint, the first equation can be factorised, so the system is

$$\dot{x} = x(1 + y - 4x^2), \qquad \dot{y} = y + 4x^2 - y^3.$$

Hence the line $x = 0$ is an invariant line, because $x = 0$ implies $\dot{x} = 0$.

To find the other two, try $y = cx$ so that $h(x, y) = y - cx = 0$. If the line is invariant, $\dot{h}$ must be zero on the line.

$$\dot{h} = \dot{y} - c\dot{x} = cx + 4x^2 - (cx)^3 - cx - c^2x^2 + 4cx^3 = (4 - c^2)x^2 + (4c - c^3)x^3.$$

This is zero when $c^2 = 4$, so the invariant lines are $x = 0$ and $y = \pm 2x$.

5.7 First factorise each equation,

$$\dot{x} = 5x - x^2 - xy = x(5 - x - y), \qquad \dot{y} = 4y - 2y^2 - xy = y(4 - 2y - x).$$

There are fixed points at $(0, 0)$, $(0, 2)$, $(5, 0)$. For the fourth fixed point, solve $5 - x - y = 0$ and $4 - 2y - x = 0$. This gives $x = 6$, $y = -1$, but this is not a valid solution for a population model, since one of the variables is negative. So there only three fixed points in the region of interest $x, y \geq 0$, making the problem qualitatively different from Example 5.8. There is no co-existence state. The Jacobian is

$$J = \begin{pmatrix} 5 - 2x - y & -x \\ -y & 4 - 4y - x \end{pmatrix},$$

which at the three fixed points evaluates to

$$J_{0,0} = \begin{pmatrix} 5 & 0 \\ 0 & 4 \end{pmatrix}, \quad J_{0,2} = \begin{pmatrix} 3 & 0 \\ -2 & -4 \end{pmatrix}, \quad J_{5,0} = \begin{pmatrix} -5 & -5 \\ 0 & -1 \end{pmatrix}.$$

Because of the zeros in the matrices we can just read off the eigenvalues λ. For $(0, 0)$, $\lambda = 5, 4$, so this is an unstable node. At $(0, 2)$, $\lambda = 3, -4$, so we have a saddle point with eigenvectors $(0, 1)$ in the stable direction and $(7, -2)$ in the unstable direction. Finally, $(5, 0)$ is a stable node. To sketch the phase plane, it is useful to find the x-nullcline, which consists of the lines $x = 0$ and $x + y = 5$, and the y-nullcline, $y = 0$ and $x + 2y = 4$, see Fig. 9.3.

The conclusion is that all trajectories, except those lying on the invariant y axis, approach the stable fixed point at $(5, 0)$, so one species becomes extinct and the other one levels out at a constant value.

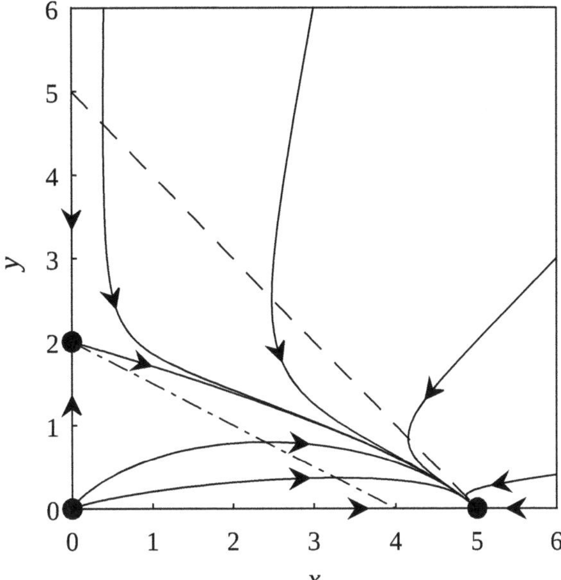

Fig. 9.3 Phase portrait for Exercise 5.7. Dashed and dash-dot lines are the x- and y-nullclines respectively

5.9 The system is

$$\dot{x} = 3x - x^2 - xy, \qquad \dot{y} = -y + xy = y(-1 + x).$$

(a) The species interaction term is $-xy$ in the x equation, $+xy$ in the y equation, so the presence of both x and y decreases the population of x and increases the population of y. Therefore y is the predator species and x is the prey. The axes $x = 0$ and $y = 0$ are both invariant lines as in the Lotka–Volterra system, but y decreases on the line $x = 0$, because the predators cannot survive if there are no prey.

(b) The fixed points are $(0, 0)$, $(3, 0)$ and $(1, 2)$. There is no predator-only steady state because of the reason given above. The Jacobian is

$$J = \begin{pmatrix} 3 - 2x - y & -x \\ y & -1 + x \end{pmatrix},$$

and evaluating J at the fixed points gives

$$J_{0,0} = \begin{pmatrix} 3 & 0 \\ 0 & -1 \end{pmatrix}, \quad J_{3,0} = \begin{pmatrix} -3 & -3 \\ 0 & 2 \end{pmatrix}, \quad J_{1,2} = \begin{pmatrix} -1 & -1 \\ 2 & 0 \end{pmatrix}.$$

The origin is a saddle point with unstable eigenvector along the x axis and stable eigenvector along the y axis. The point $(3, 0)$ is also a saddle, with its stable direction along the x axis and unstable direction $(-3, 5)$. For the point $(1, 2)$ the trace of J is $T = -1$ and the determinant is $D = 2$, so $T^2 - 4D < 0$ and this point is a stable spiral. Considering the direction of trajectories on the axes, the spiral must rotate in the anticlockwise direction. The nullclines are the axes plus the lines $y = 3 - x$ (x-nullcline) and $x = 1$ (y-nullcline).

(c) Putting all this information together gives the phase plane diagram shown in Fig. 9.4. All trajectories except those on the invariant axes approach the stable spiral at $(1, 2)$. This means that the populations of predator and prey have oscillations that decrease in amplitude and the system approaches a steady state with both species present.

5.11 For the fixed point at $(\pi, 0)$, the determinant of the Jacobian is $-g/L$, independent of β. Hence this point is always a saddle point. For the point $(0, 0)$, the determinant is $D = g/L > 0$ and the trace is $-\beta < 0$, so $T^2 - 4D = \beta^2 - 4g/L$. Referring to Table 5.1, the origin is a stable spiral if $\beta < 2\sqrt{g/L}$ and a stable node if $\beta > 2\sqrt{g/L}$. In the second case, the damping is so strong that the pendulum does not oscillate at all, it just swings down to zero. This is sometimes called the *overdamped* case. In the borderline case, $\beta = 2\sqrt{g/L}$, the origin is a stable degenerate node.

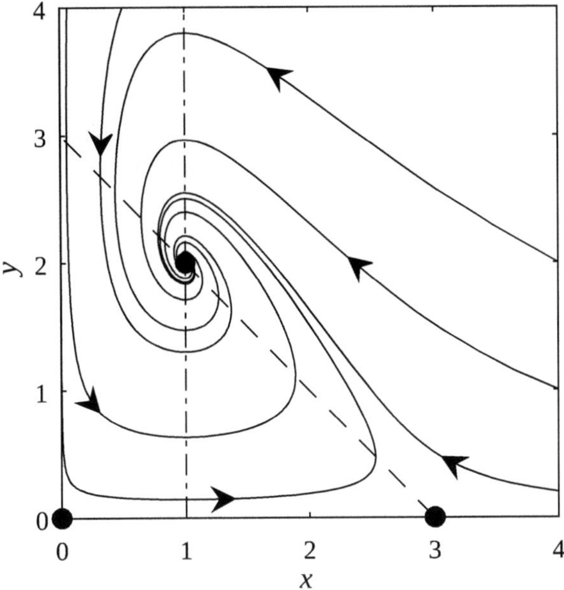

Fig. 9.4 Phase portrait for Exercise 5.9. Dashed and dash-dot lines are the x- and y-nullclines respectively

5.13 The formulas for r and θ in terms of x and y are

$$r^2 = x^2 + y^2, \qquad \theta = \tan^{-1}(y/x).$$

Differentiating each in turn,

$$2r\dot{r} = 2x\dot{x} + 2y\dot{y} \;\Rightarrow\; \dot{r} = \frac{x\dot{x} + y\dot{y}}{r},$$

$$\dot{\theta} = \frac{1}{1 + (y/x)^2}\,\frac{\dot{y}x - \dot{x}y}{x^2} = \frac{\dot{y}x - \dot{x}y}{x^2 + y^2} = \frac{\dot{y}x - \dot{x}y}{r^2},$$

which are the equations in (5.28).

5.15 For the problem of Example 5.8, the divergence of $h(x, y)(\dot{x}, \dot{y})$ is

$$\frac{\partial}{\partial x}\left(\frac{3 - x - y}{y}\right) + \frac{\partial}{\partial y}\left(\frac{4 - 2y - x}{x}\right) = -\frac{1}{y} - \frac{2}{x}.$$

In $x > 0$, $y > 0$ this is always negative, so there can be no periodic orbits.

5.17 For $\ddot{u} + (4u^2 + \dot{u}^2 - 1)\dot{u} + u = 0$, first define $v = \dot{u}$ to create a second-order system, $\dot{u} = v$, $\dot{v} = -u - (4u^2 + v^2 - 1)v$. To apply the Poincaré–Bendixson theorem, first check for fixed points. At a fixed point, $v = 0$ and $u = 0$ so the origin is the only fixed point. Now apply the polar coordinate method, using $r^2 = u^2 + v^2$.

Then

$$r\dot{r} = u\dot{u} + v\dot{v} = uv - vu - v^2(4u^2 + v^2 - 1) = -v^2(r^2 + 3u^2 - 1).$$

Since $v^2u^2 \geq 0$, $r\dot{r} \leq -v^2(r^2 - 1)$. So $\dot{r} \leq 0$ on any circle with radius greater than 1, for example the circle $r^2 = 2$. Similarly,

$$r\dot{r} = -v^2(4u^2 + v^2 - 1) = -v^2(4r^2 - 3v^2 - 1) \geq -v^2(4r^2 - 1)$$

and so $\dot{r} \geq 0$ on any circle with $4r^2 - 1 < 0$, for example the circle $r^2 = 1/5$. Therefore any trajectory that starts inside the annulus $1/\sqrt{5} < r < \sqrt{2}$ cannot leave this region, so there is a periodic orbit lying inside this annulus.

Problems of Chap. 6

6.1 For $\dot{x} = f(x, \mu) = 1 - \mu x + x^2$, bifurcations occur when $f = 0$ and $f_x = 0$ so $1 - \mu x + x^2 = 0$ and $-\mu + 2x = 0$. Eliminating $\mu = 2x$ gives $1 - x^2 = 0$ so there are two bifurcations, at $x = 1$, $\mu = 2$, and at $x = -1$, $\mu = -2$. These are also the values of μ where the number of fixed points changes from 2 to 0, so it looks as though these are saddle–node bifurcations. To confirm this, check the values of $f_{xx} = 2$ and $f_\mu = -x = \pm 1$. These are both non-zero at the bifurcation points, so both bifurcations are saddle–nodes.

6.3 For $\dot{x} = \mu^2 - x^2 = (\mu + x)(\mu - x)$, change the variable from x to $y = x - \mu$, so $\dot{y} = \dot{x} = (2\mu + y)(-y) = -2\mu y - y^2$. After relabelling $\mu = -\nu/2$ this is the normal form for a transcritical bifurcation, $\dot{y} = \nu y - y^2$. So the original equation has a transcritical bifurcation at $x = 0$, $\mu = 0$. This is as expected, because the two branches of fixed points, $x = \mu$ and $x = -\mu$, cross at $x = 0$, $\mu = 0$.

6.5

(a) For $\dot{x} = f(x, \mu) = (x^2 - 1)(\mu - x - x^2)$, fixed points are at $x = \pm 1$ or the roots of $x^2 + x - \mu = 0$. Roots exist if $\mu > -1/4$ so we expect a saddle–node bifurcation at $\mu = -1/4$. Also the quadratic crosses $x = 1$ when $\mu = 2$, and $x = -1$ when $\mu = 0$ so we expect transcritical bifurcations at these points. To check this, solve $f_x = -4x^3 - 3x^2 + 2x + 1 + 2x\mu = 0$ and $f = 0$ to find bifurcation points. Consider each solution of $f = 0$: if $x = 1$ then $f_x = -4 + 2\mu$, so there is a bifurcation at $\mu = 2$, $x = 1$, where $f_\mu = 0$ but $f_{xx} \neq 0$ and $f_{x\mu} \neq 0$ so this is a transcritical bifurcation. For $x = -1$, $f_x = -2\mu$ so there is a bifurcation at $\mu = 0$, $x = -1$ which is also transcritical. Finally, if $\mu - x - x^2$ is zero, then eliminating μ in the formula for f_x gives $(1 - x^2)(2x + 1) = 0$. So there are three bifurcations on this branch of solutions. Two of these are the transcritical bifurcations that we have already found (since transcritical bifurcations occur when two solution branches cross, they come up twice in a systematic search for all bifurcations). The third bifurcation is at $x = -1/2$, where $f_x = 0$ implies $\mu = -1/4$. At this point f_{xx} and f_μ are both non-zero, so this is a saddle–node bifurcation as predicted.

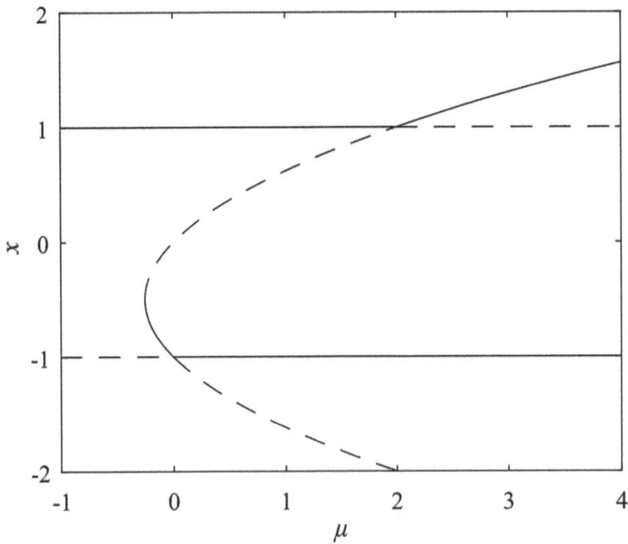

Fig. 9.5 Bifurcation diagram for Exercise 6.5. There is a saddle-node bifurcation at $\mu = -1/4$, $x = -1/2$, and transcritical bifurcations at $\mu = 0$, $x = -1$ and $\mu = 2$, $x = 1$

(b) The fixed point $x = 1$ is stable for $\mu < 2$, and unstable for $\mu > 2$, while $x = -1$ is stable for $\mu > 0$, unstable for $\mu < 0$. The stability of the quadratic curve of fixed points changes at each of the three bifurcations. The bifurcation diagram is shown in Fig. 9.5.

(c) For the initial condition $x = -1/2$, (i) If $\mu = -1$, x approaches the stable fixed point at $x = 1$. (ii) If $\mu = -3/16$, x approaches the stable fixed point where $x^2 + x - \mu = 0$, which is $x = -3/4$. (iii) If $\mu = 2$, $x \to -1$ as $t \to \infty$.

6.7 Consider $\dot{x} = \mu + \beta x - x^2$. Completing the square, this can be written as $\dot{x} = \mu - (x - \beta/2)^2 + \beta^2/4$. Now define $\hat{x} = x - \beta/2$ and $\hat{\mu} = \mu + \beta^2/4$, so that $\dot{\hat{x}} = \hat{\mu} - \hat{x}^2$, which is equivalent to (6.5).

6.9

(a) For $\dot{x} = y$, $\dot{y} = -y - x^2$, the origin is a fixed point. The linearised form is

$$\begin{pmatrix} \dot{x} \\ \dot{y} \end{pmatrix} = \begin{pmatrix} 0 & 1 \\ 0 & -1 \end{pmatrix} \begin{pmatrix} x \\ y \end{pmatrix},$$

so there is a zero eigenvalue with eigenvector in the x direction.

(b) In the expansion $y = a + bx + cx^2 + dx^3 + O(x^4)$, the centre manifold must go through the origin, so $a = 0$, and the x axis must be a tangent, so $b = 0$.

(c) The curve $h(x, y) = y - cx^2 - dx^3 = 0$ must be invariant, so on this curve, $\dot{h} = \dot{y} - (2cx + 3dx^2)\dot{x} = 0$. Substituting for $\dot{x}$ and $\dot{y}$, $-y - x^2 - (2cx + 3dx^2)y =$

0. Now substitute in the expansion for y, keeping only the terms up to x^3. This gives $-cx^2 - dx^3 - x^2 - 2c^2x^3 + O(x^4) = 0$. From the x^2 terms, $c = -1$, and the x^3 terms give $d = -2c^2 = -2$, so the CM is $y = -x^2 - 2x^3 + O(x^4)$.

(d) Near the origin, solutions approach the centre manifold curve and move along it according to $\dot{x} = y = -x^2 - 2x^3 + O(x^4)$. If $x > 0$ then $x \to 0$, but if $x < 0$, x moves away from 0, so the origin is unstable.

6.11 Adding a μx term to the $\dot{x}$ equation in Exercise 6.10 gives $\dot{x} = \mu x + x^2 y - 2xy^2$, $\dot{y} = x - y + 2x^2 y$.

(a) The origin is a fixed point, for any μ. The Jacobian is as in Exercise 6.10 but with $J_{1,1} = \mu$, so the determinant is $-\mu$. Hence the origin has a stationary bifurcation at $\mu = 0$, using Definition 6.2.

(b) The given ECM is $y = x + a\mu + b\mu x + c\mu^2 x + d\mu x^2 + ex^3 + O(4)$. This must go through the fixed point $(0, 0)$, so $a = 0$. Also, the rotation symmetry $(x, y) \leftrightarrow (-x, -y)$, which is retained with the additional μx term, means that the ECM must be an odd function of x, so $d = 0$. When $\mu = 0$, from Exercise 6.10, $e = 3$.

(c) Now apply the invariant curve condition, which is

$$\dot{y} - (1 + b\mu + c\mu^2 + 3ex^2)\dot{x} = 0,$$

and substituting for $\dot{y}$ and $\dot{x}$ gives

$$x - y + 2x^2 y - (1 + b\mu + c\mu^2 + 3ex^2)(\mu x + x^2 y - 2xy^2) = 0.$$

Substituting in the ECM for y, with $a = d = 0$, and equating coefficients of x^3, μx and $\mu^2 x$ confirms that $e = 3$ and gives $b = -1$, $c = 1$, so the ECM is $y = x - \mu x + \mu^2 x + 3x^3 + O(4)$.

(d) Substituting the ECM into the $\dot{x}$ equation and taking only terms up to third order in x and μ gives $\dot{x} = \mu x - x^3$, so the system has a supercritical pitchfork bifurcation at $\mu = 0$. As in some other examples, the computed coefficients b and c do not influence the outcome, so we have included more terms than necessary. Also, a quick way to find the type of bifurcation is to use the rotation symmetry to deduce that the bifurcation is a pitchfork, and then use the information from the previous solution that the origin is stable when $\mu = 0$ to see that the pitchfork must be supercritical.

6.13 In Example 6.7 the fixed points are at the origin and $x = \pm\sqrt{\mu}$, $y = -2\mu$ (if $\mu > 0$). The Jacobian matrix at a general point and at the fixed points is

$$J = \begin{pmatrix} \mu + 3x^2 + y & x \\ -4x & -1 \end{pmatrix}, \quad J_{0,0} = \begin{pmatrix} \mu & 0 \\ 0 & -1 \end{pmatrix}, \quad J_{\pm\sqrt{\mu},-2\mu} = \begin{pmatrix} 2\mu & \pm\sqrt{\mu} \\ \mp4\sqrt{\mu} & -1 \end{pmatrix}.$$

At the origin the eigenvalues are always real, $\lambda = \mu$ and $\lambda = -1$, so there is no Hopf bifurcation. For the other fixed points the trace of J is $T = 2\mu - 1$ and the determinant is $2\mu > 0$. There is a Hopf bifurcation when the eigenvalues are imaginary, $\lambda = \pm i\omega$, which happens when $T = 0$, so $\mu = 1/2$.

6.15 There is a fixed point at $(0, 0)$ and the linearised system is

$$\dot{x} = \mu x - y, \quad \dot{y} = x + \mu y,$$

with eigenvalues $\lambda = \mu \pm i$, so the origin is stable for $\mu < 0$ and there is a Hopf bifurcation at $\mu = 0$. Translating the system into polar coordinates (see Example 5.11) leads to

$$\dot{r} = \mu r + r^3 - r^5, \quad \dot{\theta} = 1 - r^2.$$

Since $\dot{r} = \mu r + r^3 + O(r^5)$, the Hopf bifurcation is subcritical. There are periodic orbits when $\dot{r} = 0$, so $r^4 - r^2 - \mu = 0$. This has no solutions for $\mu < -1/4$ and two for $\mu > 1/4$, with a saddle–node bifurcation of periodic orbits at $\mu = 1/4$. However, r^2 must be positive, which is only true for both roots if $\mu < 0$. In summary, there are two periodic orbits for $-1/4 < \mu < 0$, one stable and one unstable, and one stable periodic orbit for $\mu > 0$. The bifurcation diagram is the same as that shown in Exercise 6.6, except that the fixed points in that example are periodic orbits in this question. The orbit rotates clockwise if $r^2 > 1$, so $\mu = r^2(r^2 - 1) > 0$. For $-1/4 < \mu < 0$, both orbits go anticlockwise. When $\mu = 0$, the fixed point of the r equation is $r = 1$, which is also a fixed point of the θ equation, so there is a circle of fixed points.

Problems of Chap. 7

7.1

(a) Set $x_n = Ap^n$ in (7.2). Then $Ap^{n+2} = Ap^{n+1} + Ap^n$ and so $p^2 - p - 1 = 0$. Hence $p = (1 \pm \sqrt{5})/2$. Set $p_+ = (1 + \sqrt{5})/2$ and $p_- = (1 - \sqrt{5})/2$. Because (7.2) is linear and homogeneous, we can use a superposition principle very similar to that of Sect. 2.4, and the general solution is

$$x_n = Ap_+^n + Bp_-^n,$$

for any constants A and B.

(b) If $x_0 = x_1 = 1$ then $A + B = 1$ and $Ap_+ + Bp_- = 1$, and solving these gives $A = p_+/\sqrt{5}$, $B = -p_-/\sqrt{5}$, so

$$x_n = (p_+^{n+1} - p_-^{n+1})/\sqrt{5}.$$

Using this formula, $x_2 = (p_+^3 - p_-^3)/\sqrt{5} = (p_+ - p_-)(p_+^2 + p_+ p_- + p_-^2)/\sqrt{5} = (1 + 5 + 1 - 5 + 1 + 5)/4 = 2$, which is correct. Although the formula involves $\sqrt{5}$, the result never does!

(c) As $n \to \infty$, $x_{n+1}/x_n \to p_+ = 1.618\ldots$, which is known as the Golden Ratio.

7.3

(a) The quadratic in Exercise 7.1 is $x^2 - x - 1 = 0$. Halley's method is

$$x_{n+1} = x_n - \frac{2(x_n^2 - x_n - 1)(2x_n - 1)}{2(2x_n - 1)^2 - 2(x_n^2 - x_n - 1)}.$$

Starting from $x_0 = 2$, $x_1 = 2 - 3/8 = 1.625$ and $x_2 = 1.618034$ to six decimal places.

(b) For Newton's method, $x_{n+1} = x_n - (x_n^2 - x_n - 1)/(2x_n - 1)$, $x_0 = 2$, $x_1 = 2 - 1/3 = 1.666667$ and $x_2 = 1.619048$. To six places, the true solution, p_+ in Exercise 7.1, is 1.618034, so Halley's method reaches this level of accuracy in just two steps, and converges faster than Newton's method.

7.5

(a) The differential equation $\dot{x} = x - x^2$ is separable and the solution is $x = 1/(1 + Ce^{-t})$ (see Exercise 2.2(a), which is almost identical). With the initial condition $x(0) = 2$, $C = -0.5$.

(b) Adapting the code to print the error at each step shows that with $h = 0.2$, the maximum error is $0.096\ldots$, occurring at the second step, corresponding to $t = 0.4$.

7.7 The difference equation factorises to give

$$x_{n+1} - x_n = f(x_n) = (x_n^2 - 1)(x_n^2 + x_n/2) = (x_n + 1)(x_n - 1)(x_n)(x_n + 1/2),$$

so the fixed points are $x_* = -1, -1/2, 0, 1$. The stability of each point depends on $f'(x_*) = 4x_*^3 + 3x_*^2/2 - 2x_* - 1/2$. At the four fixed points, $f'(-1) = -1$ (superstable), $f'(-1/2) = 3/8$ (unstable), $f'(0) = -1/2$ (stable), $f'(1) = 3$ (unstable).

7.9

(a) For $f(x, \mu) = e^x - x + \mu$, there is a fixed point where $\mu = x - e^x$. There is a stationary bifurcation if $f_x = e^x - 1 = 0$, so $x = 0$ and hence $\mu = -1$. Since $f_{xx} = e^x \neq 0$ and $f_\mu = 1 \neq 0$, the bifurcation is a saddle–node. For small x, $f(x, \mu) = 1 + \mu + x^2/2 + O(x^3)$, which is another way to show that the bifurcation is a saddle–node. The two fixed points exist for $\mu < -1$.

(b) For $f(x, \mu) = x/(1 + x^2) - \mu x$, $x = 0$ is a fixed point for all μ. Since $f_x = (1 - x^2)/(1 + x^2)^2 - \mu = 1 - \mu$ at $x = 0$, the fixed point is stable for $1 < \mu < 3$, and there is a stationary bifurcation at $\mu = 1$, $x = 0$. There is a non-zero fixed point at $\mu = 1/(1 + x^2)$, which exists for $0 < \mu < 1$, where the zero fixed point is unstable, so there is a supercritical pitchfork bifurcation at $\mu = 1$.

7.11

(a) Newton's method for $G(x) = x^3 - x + \mu = 0$ is

$$x_{n+1} = x_n - \frac{x_n^3 - x_n + \mu}{3x_n^2 - 1}.$$

(b) If $x_0 = 0$ then $x_1 = \mu$ and $x_2 = \mu - \mu^3/(3\mu^2 - 1) = (2\mu^3 - \mu)/(3\mu^2 - 1)$. There is a period-two cycle if $x_2 = x_0 = 0$, so $\mu = 1/\sqrt{2}$.

(c) From (7.12) and (7.23), a period-two cycle of $x_{n+1} = x_n + f(x_n)$ is superstable if $f'(x_0) + f'(x_1) + f'(x_0)f'(x_1) = -1$. This is true if either $f'(x_0) = -1$ or $f'(x_1) = -1$. Also, from Example 7.3, $f'(x) = G(x)G''(x)/G'(x)^2 - 1$, which is -1 if either $G(x) = 0$ as in that example, or $G''(x) = 0$ which applies here, since $G''(x) = 6x = 0$ at $x = x_0 = 0$. So for $\mu = 1/\sqrt{2}$ there is a cycle $(0, \mu)$ that is superstable.

7.13

(a) If x_* is a fixed point of the differential equation, then $f(x_*) = 0$. If $X_n = x_*$, $X_{n+1} = x_* + hf(x_* + hf(x_*)/2) = x_* + hf(x_*) = x_*$, so $X_{n+1} = X_n = x_*$ and we have a fixed point of the difference equation.

(b) The formula (7.36) is $X_{n+1} = X_n + hf(z)$, where $z = X_n + hf(X_n)/2$. So at a fixed point, $f(z) = 0$. For $f(x) = x(1 - x)$, fixed points X_* are where $z = 0$ or $z = 1$. If $z = 0$ then $X_* + hX_*(1 - X_*)/2 = 0$, so either $X_* = 0$ or $X_* = 1 + 2/h$. If $z = 1$ then $X_* + hX_*(1 - X_*)/2 = 1$, so either $X_* = 1$ or $X_* = 2/h$. Hence the difference equation has four fixed points, two that correspond to the fixed points of the differential equation and two 'spurious' ones that depend on h.

(c) The fixed point $X_* = 1$ is stable if $-2 < hf'(X_* + hf(X_*)/2)(1 + hf'(X_*)/2) < 0$. Since $f(X_*) = 0$ and $f'(X_*) = 1 - 2X_* = -1$, we have stability if $-2 < -h(1 - h/2) < 0$. The inequality on the left is always true. The one on the right is true if $h < 2$. The stability condition on h is exactly the same as for Euler's method, see Exercise 7.8(b). But the type of instability is different because the other stability condition is broken. Euler's method has a period-doubling bifurcation, but the midpoint method has a stationary bifurcation at $h = 2$.

(d) There are several ways of seeing that the bifurcation at $h = 2$ is transcritical.

(i) The fixed point $X_* = 1$ exists for all h, which rules out a saddle-node bifurcation, and there is no symmetry, so it is unlikely to be a pitchfork.

(ii) The spurious solution $X_* = 2/h$ crosses the true solution $X_* = 1$ at $h = 2$, indicating a transcritical bifurcation.

(iii) We could expand the method near $x = 1$, $h = 2$ to derive the normal form (7.15).

Problems of Chap. 8

8.1 Using the trig rules for products of functions (see Sect. A.2.3),

$$S_m = \int_0^{2\pi} \sin\theta \, \sin(m\theta) \, d\theta = \frac{1}{2} \int_0^{2\pi} \cos(1-m)\theta - \cos(1+m)\theta \, d\theta \,.$$

For $m = 1$,

$$S_1 = \frac{1}{2} \int_0^{2\pi} 1 - \cos 2\theta = \frac{1}{2} \left[\theta - \frac{1}{2} \sin 2\theta \right]_0^{2\pi} = \frac{1}{2}(2\pi) = \pi.$$

For all other integer values of m,

$$S_m = \frac{1}{2} \left[\frac{1}{1-m} \sin(1-m)\theta - \frac{1}{1+m} \sin(1+m)\theta \right]_0^{2\pi} = 0.$$

Similarly,

$$C_m = \int_0^{2\pi} \sin\theta \, \cos(m\theta) \, d\theta = \frac{1}{2} \int_0^{2\pi} \sin(1-m)\theta + \sin(1+m)\theta \, d\theta \,.$$

Hence

$$C_1 = \frac{1}{2} \int_0^{2\pi} \sin 2\theta \, d\theta = \frac{1}{4} [-\cos 2\theta]_0^{2\pi} = 0,$$

and for $m \neq 1$,

$$C_m = \frac{1}{2} \left[-\frac{1}{1-m} \cos(1-m)\theta - \frac{1}{1+m} \cos(1+m)\theta \right]_0^{2\pi} = 0.$$

So after substituting the series (8.3) into (8.1), there are infinitely many integrals, but they are all zero except for the one coming from the $b_1 \sin\theta$ term, which gives the result (8.4).

8.3 Let $V(x, y, z) = \frac{r}{\sigma}x^2 + y^2 + (z - 2r)^2$. Then the surfaces $V = $ constant are ellipsoids with centre at $(0, 0, 2r)$. Following the method of Sect. 8.1.2,

$$\dot{V}/2 = rx(y - x) + y(rx - xz - y) + (z - 2r)(xy - z)$$

$$= -rx^2 - y^2 - z^2 + 2rz = -rx^2 - y^2 - (z - r)^2 + r^2.$$

Therefore $\dot{V} < 0$ outside the ellipsoid E_0 defined by $rx^2 + y^2 + (z-r)^2 = r^2$ (which is *not* one of the $V = c$ ellipsoids). Now choose the constant c to be large enough so that the $V = c$ ellipsoid completely contains the ellipsoid E_0. Then $\dot{V} < 0$ on

$V = c$, so all trajectories point inward on this surface, and trajectories of the Lorenz equations are bounded by it.

8.5 Equation (8.14) gives the value of r at which the Hopf bifurcation occurs as a function of σ, provided that $\sigma > 2$. It is clear from the formula that $r_H \to \infty$ as $\sigma \to \infty$ and as $\sigma \to 2$, so there must be a minimum at some value of σ. Setting the derivative to 0, the minimum occurs when $(2\sigma + 4)(\sigma - 2) - \sigma(\sigma + 4) = 0$, so $\sigma^2 - 4\sigma - 8 = 0$ and hence $\sigma = 2 + \sqrt{12} = 2(1 + \sqrt{3}) \approx 5.464$. The corresponding minimum value of r_H is $4(2 + \sqrt{3}) \approx 14.928$.

8.7

(a) For $\sigma = 5$, $r = 5$, the calculations of Sect. 8.1.3 show that the non-zero fixed points $(2, 2, 4)$ and $(-2, -2, 4)$ are stable. The computer code should show that x, y and z oscillate and approach one of these points, depending on the initial conditions. A phase space plot shows the solution spiralling inwards. If Euler's method is used and the step size is increased to $h = 0.1$, the code incorrectly shows what appears to be a chaotic solution, and if h is greater than about 0.2 (depending on your choice of initial conditions) the variables blow up and the code crashes.

(b) For $\sigma = 5$, $r_H = 15$, so for $r = 14$ and initial conditions close to $(\sqrt{13}, \sqrt{13}, 13)$, trajectories should spiral in towards the fixed point. With Euler's method, this happens if h is less than about 0.0024. For larger h, trajectories incorrectly spiral outwards. For $r = 16$, solutions spiral out from the fixed point and become chaotic.

(c) Choose $h = 0.001$. For $r = 14 < r_H$, solutions with initial conditions close to the fixed point spiral into it, as found in (b). Initial conditions close to the origin also lead to a stable fixed point. But other choices give chaotic solutions. This is to be expected because the Hopf bifurcation is subcritical. For $r = 13$, *transient chaos* can be found. The solution behaves chaotically for a finite time, typically 50–100 time units, and then spirals in to a stable fixed point.

(d) You will not be able to reproduce Fig. 8.2 exactly, because small numerical errors are amplified exponentially by the chaotic system. Your graph will probably only match Fig. 8.2 for the first 10–20 time units.

8.9 For general σ, the Hopf bifurcation is at $r = \sigma(\sigma + 4)/(\sigma - 2)$, and $\omega^2 = \sigma + r$. Following the method of Example 8.1, the two complex eigenvalues are $\pm i\omega$. The complex eigenvectors can be written as $(x_0\sigma, x_0(\sigma \pm i\omega), \omega^2 \mp i\omega(1 + \sigma))$, where $x_0 = \sqrt{r - 1}$. Hence two real vectors in the plane are $(x_0\sigma, x_0\sigma, \omega^2)$ and $(0, \omega x_0, -(1 + \sigma)\omega)$. Taking the cross product of these and taking out factors to simplify the result, a vector normal to the surface is $(-\sigma/(\sigma - 2), 1, x_0/(1 + \sigma))$. The equation of the plane tangent to the centre manifold is $-\sigma x/(\sigma - 2) + y + x_0 z/(1 + \sigma) = $ constant. For $\sigma = 5$, the normal vector is $(-5/3, 1, \sqrt{14}/6)$ which is a multiple of the vector given in Example 8.1.

8.11 All four of these difference equations have bifurcation diagrams similar to those shown in this chapter, with a sequence of period doubling bifurcations, a region of chaos, and windows of periodic behaviour.

8.13 From Exercise 8.12, a period four solution is stable at $\mu = 1.6$, followed by period eight near $\mu = 1.63$ and period 16 around $\mu = 1.645$. Within the chaotic region there are infinitely many windows of stable periodic behaviour! Some of these are, in order of increasing μ:

Period:	12	10	6	5	7
μ:	1.667	1.697	1.726	1.875	1.924

8.15

(a) The method is the same as for Example 8.3, so only key points in the calculation are given here. Substitute $x_n = a + b\omega^n + \bar{b}\bar{\omega}^n$ into $(x_{n+1} - x_n)(1 + x_n) = -\mu x_n$ and multiply out, using the properties (8.19). Then equate the constant terms, ω^n terms and $\bar{\omega}^n$ terms to get

$$\mu a = 3|b|^2,$$
$$b(\omega - 1)(1 + a) + \mu b = \bar{b}^2(1 - \bar{\omega}),$$
$$\bar{b}(\bar{\omega} - 1)(1 + a) + \mu \bar{b} = b^2(\omega - 1).$$

Combining the second and third equations gives a formula for $|b|^2$ in terms of a, and using the first equation to eliminate $|b|^2$ gives a quadratic for a,

$$3a^2 + (6 - 4\mu)a + \mu^2 - 3\mu + 3 = 0.$$

This has real solutions for $\mu > 3$, so this is where period-three solutions exist.

(b) At the saddle-node bifurcation, $\mu = 3$, the equation for a is $a^2 - 2a + 1 = 0$, so $a = 1$ and hence $|b|^2 = 1$. The second of the three equations above gives

$$b^3(2(\omega - 1) + 3) = 1 - \bar{\omega}$$

so $b^3 = (1 - \sqrt{3}i)/2 = e^{-i\pi/3}$ and choosing the root $b = e^{-i\pi/9}$, the period-three cycle consists of the points

$$x_0 = 1 + 2\cos(\pi/9), \quad x_1 = 1 + 2\cos(5\pi/9), \quad x_2 = 1 + 2\cos(11\pi/9).$$

8.17 With a random initial condition, a computer code for the tent map shows chaotic solutions initially, but after about 60 iterations, the solution tends to the fixed point $x = 0$, which ought to be unstable. Similarly, with the initial condition $x_0 = 1/7$, the solution follows the period-three cycle of Example 8.4 for a while, but again, after about 60 steps the iterations reach zero. This is because of the limited precision of numbers on a computer. Most systems use 64-bit real numbers, meaning 64 binary digits. After one iteration, since the tent map involves a multiplication by 2, the final digit stored must be a zero. After two iterations, the last two binary digits are zero, and so on, until all digits are zero.

8.19 Computing 50 iterations of the tent map from the initial condition $x_0 = 0.23$ shows a period-ten cycle, although when this is continued to 70 iterations the solution drops to 0, for the same reason as in Exercise 8.17. Initial conditions 0.51, 0.19 and 0.78 all show the same behaviour.

The initial conditions have the form $x_0 = m/100$. If m is odd, and not a multiple of 5, then x_3 must be of the form $p/25$, where p is even. Numbers of this form where p is not a multiple of 5 form a cycle under the tent map: $(2, 4, 8, 16, 18, 14, 22, 6, 12, 24)/25$, which has length 10. If m is even, this cycle is reached sooner. There are two special cases. If $m = 25, 50$ or 75 then $x_j = 0$ for $j > 2$. For other multiples of 5, $m = 5, 10, 15..., x_j$ lies on the two-cycle $(2/5, 4/5)$ for $j > 2$.

8.21 Using point 3 of Sect. 8.3.6, for a period-five orbit, $2^5 x_0 = m + x_0$, so $x_0 = m/31$ and $q = 31$. The distinct period-five solutions are generated by 5-bit binary numbers that are not related by a bit shift: 00001, 00011, 00101, 00111, 01011, 01111. Hence there are six period-five cycles, which agrees with the formula from point 4, since $(2^5 - 2)/5 = 6$.

8.23 For $x_{n+1} = \mu x_n \bmod 1$, with $1 \leq \mu \leq 2$, the alternative form is

$$
x_{n+1} = \begin{cases} \mu x_n & \text{for } 0 \leq x_n < 1/\mu, \\ \mu x_n - 1 & \text{for } 1/\mu \leq x_n \leq 1. \end{cases}
$$

(a) Consider a period-two cycle with $x_0 < 1/\mu$, so $x_1 = \mu x_0 > x_0$. If $x_1 < 1/\mu$ then $x_2 = \mu^2 x_0 \neq x_0$ so there cannot be a two-cycle. So $x_1 > 1/\mu$ and $x_2 = \mu x_1 - 1$, and there is a period-two orbit if $x_0 = \mu^2 x_0 - 1$, leading to $x_0 = 1/(\mu^2 - 1)$, $x_1 = \mu/(\mu^2 - 1)$. For this to be consistent, $x_0 < 1/\mu \Rightarrow \mu^2 - \mu - 1 > 0$, so as in Exercise 8.20, $\mu > \phi = (1 + \sqrt{5})/2 = 1.618....$
(b) For a period-three cycle of type LLR, $x_1 = \mu x_0$, $x_2 = \mu x_1$, $x_3 = \mu x_2 - 1 = \mu^3 x_0 - 1 = x_0$, so $x_0 = 1/(\mu^3 - 1)$, $x_1 = \mu/(\mu^3 - 1)$, $x_2 = \mu^2/(\mu^3 - 1)$. Consistency requires $x_1 < 1/\mu$, so $\mu^3 - \mu^2 - 1 > 0$, and $x_2 < 1$ which gives the same cubic inequality. Using the hint, this inequality is true, so a period-three solution exists, if $\mu > 1.46....$
(c) The results of (a) and (b) seem to contradict Sharkovsky's theorem. For $\mu = 1.5$, a period-three solution exists (explicitly, it is $[8/19, 12/19, 18/19]$), but no period-two solution exists, since $\mu < \phi$. However, Sharkovsky's theorem requires x_{n+1} to be a continuous function of x_n, which is not the case here, so the theorem cannot be applied and there is no contradiction.

Appendix A
Essential Background Mathematics

This appendix includes summaries of a few topics that are important background knowledge for this book: order notation, functions, curve sketching, eigenvalues and eigenvectors, partial differentiation, and vector calculus.

A.1 Order Notation

Order notation, or "O" notation, is a way of keeping track of error terms when making an approximation or when using a truncated series. This notation replaces "$\approx$" or "$+\ldots$" with something more precise. The definition is

$$f(x) = O(g(x)) \text{ as } x \to 0 \quad \Leftrightarrow \quad \lim_{x \to 0} \frac{f(x)}{g(x)} = C,$$

where C is some non-zero constant (some books allow C to be zero; there are pros and cons of each version). In most applications, $g(x)$ is a power of x. Examples include

$$\sin x = O(x), \quad \sin x = x + O(x^3), \quad f(a+h) = f(a) + hf'(a) + O(h^2).$$

In the second example a $+$ sign is used even though the next term is negative. The order notation is only concerned with the magnitude of the remainder, not its sign. Although most commonly used in the limit that a variable tends to zero, the O notation can also be used for the limit $x \to \infty$ or any finite limit. For example,

$$\frac{3x+2}{1+x^2} = O\left(\frac{1}{x}\right) \text{ as } x \to \infty, \quad \sin x - 1 = O\left(x - \frac{\pi}{2}\right)^2 \text{ as } x \to \frac{\pi}{2}.$$

P. C. Matthews, *Differential Equations, Bifurcations and Chaos*,
Springer Undergraduate Mathematics Series,
https://doi.org/10.1007/978-3-031-99543-9

When there are two small parameters, say x and y, the notation $O(2)$ can be used to mean all quadratic terms, $O(x^2) + O(xy) + O(y^2)$.

A.2   Exponentials, Logs and Trig Functions

This section summarises some of the properties of the most important mathematical functions used in this book. These functions are plotted in Fig. A.1.

A.2.1   The Exponential Function

The exponential function may be defined as the function $y(x)$ that obeys the differential equation $y' = y$ and $y(0) = 1$, or via the series

$$e^x = \sum_{n=0}^{\infty} \frac{x^n}{n!} = 1 + x + \frac{x^2}{2!} + \frac{x^3}{3!} + O(x^4). \tag{A.1}$$

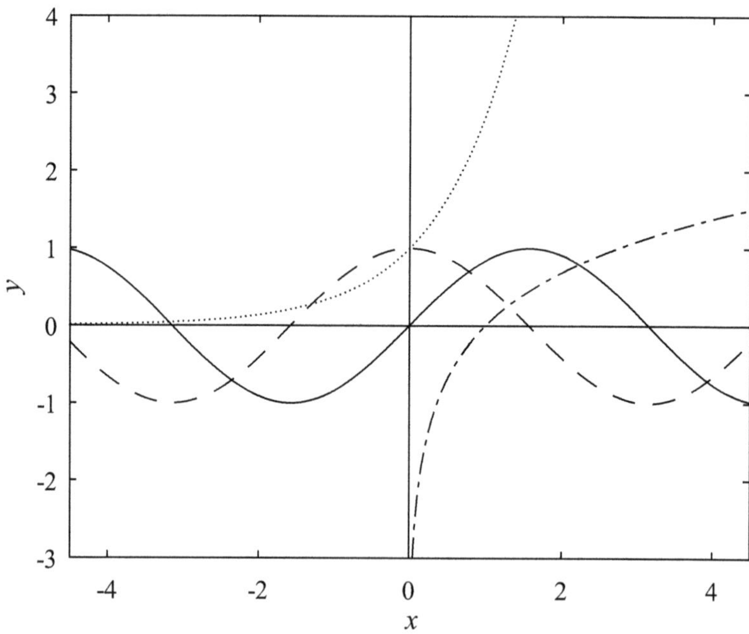

Fig. A.1 The functions e^x (dotted line), $\log x$ (dash-dot line), $\sin x$ (solid line) and $\cos x$ (dashed line)

Properties include:

$$e^x e^y = e^{x+y}, \qquad (e^x)^p = e^{px}, \qquad \frac{d}{dx}(e^{ax}) = a e^{ax}, \qquad e^x > 0.$$

A.2.2 The Logarithm Function

The logarithm is the inverse of the exponential, so if $y = e^x$ then $x = \log y$. Hence $x = \log(e^x)$. In this book, all logarithms are logs to base e. $\log x$ is only defined as a real number if $x > 0$. Properties include:

$$\log(xy) = \log x + \log y, \qquad \log(x^p) = p \log x, \qquad \frac{d}{dx}(\log x) = \frac{1}{x}. \qquad \text{(A.2)}$$

Many of the properties of the logarithm can be deduced directly from the properties of the exponential. A useful series is

$$\log(1 + x) = \sum_{n=1}^{\infty} (-1)^{n+1} \frac{x^n}{n} = x - \frac{x^2}{2} + \frac{x^3}{3} - \frac{x^4}{4} + O(x^5).$$

A.2.3 Trigonometric Functions

The trigonometric functions are related to the exponential by Euler's formula,

$$e^{ix} = \cos x + i \sin x, \qquad \text{(A.3)}$$

where $i^2 = -1$. The series for these functions are

$$\sin x = x - \frac{x^3}{3!} + \frac{x^5}{5!} + O(x^7), \qquad \cos x = 1 - \frac{x^2}{2!} + \frac{x^4}{4!} + O(x^6).$$

If you forget these, you can deduce them from (A.3) and (A.1). $\sin x$ is an odd function, meaning that $\sin(-x) = -\sin(x)$ and $\cos x$ is even, so $\cos(-x) = \cos(x)$.

Other important properties of the trig functions are

$$\frac{d}{dx}(\sin(ax)) = a \cos(ax), \qquad \frac{d}{dx}(\cos(ax)) = -a \sin(ax),$$

$$\sin(2x) = 2 \sin x \cos x, \qquad \cos(2x) = \cos^2 x - \sin^2 x, \qquad \sin^2 x + \cos^2 x = 1,$$

$$2 \sin a \cos b = \sin(a+b) + \sin(a-b), \qquad 2 \sin a \sin b = \cos(a-b) - \cos(a+b),$$

$$\sin(x + \pi/2) = \cos(x), \qquad \cos(x + \pi/2) = -\sin(x).$$

If n is an integer then

$$\sin(n\pi) = 0, \qquad \cos(n\pi) = (-1)^n, \qquad \sin(x + 2n\pi) = \sin(x).$$

A.3   Graph Sketching

Sketching curves is important in many areas of mathematics, but particularly in this subject, where the main aim is to sketch solutions of a differential equation without actually solving it. Unfortunately, graph sketching is rarely explicitly taught in undergraduate courses. There is no simple recipe to follow that will work for all functions.

Table A.1 lists some curve-sketching techniques, with some examples. The following example uses some of these methods.

Example A.1   Sketch the function

$$f(x) = \frac{x^2 + x - 2}{x^2 - x - 2} = \frac{(x+2)(x-1)}{(x+1)(x-2)}.$$

When $x = 0$, $f(x) = 1$. For small x, we can expand using the binomial expansion $(x^2 - x - 2)^{-1} = (-1/2)(1 + x/2 - x^2/2)^{-1} = (-1/2)(1 - x/2 + O(x^2))$, so $f(x) = 1 - x + O(x^2)$. Hence $y = f(x)$ crosses $x = 0$ at $y = 1$ with slope -1.

For large $|x|$, the largest terms in the numerator and denominator are both x^2, so $f(x) \to 1$ as $x \to \pm\infty$. To check if $f(x) \to 1$ from above or below, we can write

$$f(x) = \frac{x^2 + x - 2}{x^2 - x - 2} = \frac{x^2 - x - 2 + 2x}{x^2 - x - 2} = 1 + \frac{2x}{x^2 - x - 2}.$$

When x is large and positive, the extra term added to 1 is small and positive, so the approach to 1 is from above, but if x is large and negative, the extra term is negative, so $f(x)$ tends to 1 from below as $x \to -\infty$.

Near $x = -1$, $f(x) \approx 2/(3(x+1))$ so $f(x) \to -\infty$ as $x \to -1$ from below and $+\infty$ as $x \to -1$ from above. Similarly, near $x = 2$, $f(x) \approx 4/(3(x-2))$ which tends to $-\infty$ as $x \to 2$ from below and $+\infty$ as $x \to 2$ from above.

Finally, $f(x)$ has roots where $(x+2)(x-1) = 0$, so $x = -2$ and $x = 1$. Looking at the sign of $f(x)$, we can see that the function decreases from positive to negative values near each of these roots.

Using all this information, we can sketch the curve of $f(x)$, see Fig. A.2.

Table A.1 Techniques for sketching a function $f(x)$

Method	Examples
What happens when x is small? Use known approximations or Taylor series	$x \sin x \approx x^2$, $\sin x / x \approx 1$, $e^x - 1 \approx x$, $x^3 - 4x^2 - 2x + 1 \approx 1 - 2x$, $\frac{1+2x}{1-x} \approx 1$
What happens when x is large? Look at largest powers of x	$\frac{1+2x}{1-x} \approx -2$, $\sin x + x + x^2 \approx x^2$, $\frac{1+3x^2}{1-x^2+x^4} \approx 3/x^2$, $e^{-1/x} \approx 1$
Is $f(x)$ even? True if $f(x) = f(-x)$. Mirror symmetry $x \leftrightarrow -x$. $f'(0) = 0$	x^n if n is even, $x^4 + 3x^2 + 2$, $\cos x$, $\cosh x$, odd function $\times$ odd function, $x \sin x$, $\sin 2x \tan x$, $\sin(x^2)$, $\cos(x^3)$
Is $f(x)$ odd? True if $f(x) = -f(-x)$. Rotation symmetry about the origin. $f(0) = 0$	x^n if n is odd, $x - x^3$, $\sin x$, $\tan x$, $\sinh x$, even function $\times$ odd function, $x \cos x$, $x^2 \sinh x$
Have a 'library' of standard functions that you know how to sketch	$mx + c$, x^2, x^3, x^n, $\sin x$, $\cos x$, e^x, e^{-x}, $\log x$, $\sinh x$, $\cosh x$
Differentiate, to find the sign of $f'(x)$	$e^{-1/x}$: derivative is always > 0
Where $f'(x) = 0$, $f(x)$ has a stationary point, max if $f'' < 0$, min if $f'' > 0$	$x^4 - 2x^2 + 3$: max at $x = 0$, min at ± 1
Does the function blow up? If so, where? Each side of the blow-up point, is f positive or negative?	$\frac{1}{4-x^2}$ blows up at $x = \pm 2$, positive for $-2 < x < 2$, negative for $|x| > 2$. $\frac{1}{(4-x)^2}$ blows up at $x = 4$, +ve both sides
Shifting: $f(x + a)$ looks the same as $f(x)$ but shifted to the left a distance a	$(x + 4)^2$ looks like x^2 but with the minimum shifted to $x = -4$
Stretching: $f(ax)$ looks like $f(x)$ but stretched if $a < 0$, squashed if $a > 0$	$\sin(x/5)$ is zero at $x = 0, 5\pi, 10\pi \ldots$
Where does the function have roots?	$x - x^3$ has three roots, $x = -1, 0, 1$, $\log x$ has only one root, at $x = 1$, $\sin(x^2)$ has roots at $x = \sqrt{n\pi}$

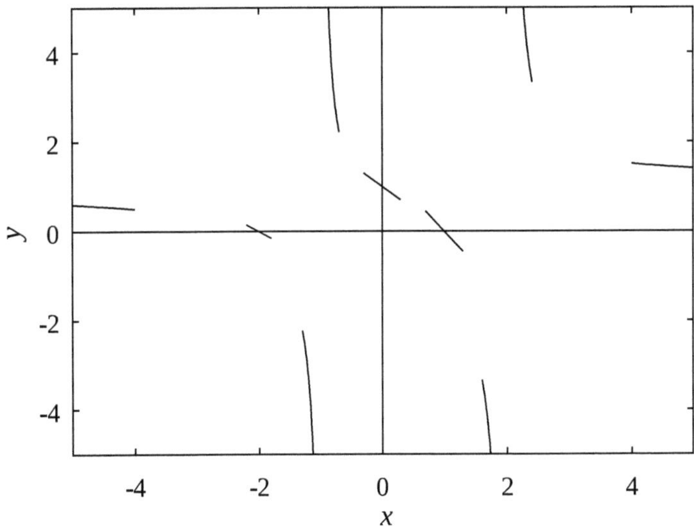

Fig. A.2 Key sections of the function $f(x)$ in Example A.1. Joining these sections with a smooth curve gives a good sketch of the function

A.4   Eigenvalues and Eigenvectors

Definition A.1 Let M be a square matrix. If there exists a non-zero vector v and a real or complex scalar λ such that

$$Mv = \lambda v \tag{A.4}$$

then λ is said to be an *eigenvalue* of M with *eigenvector v*.

Example A.2 Consider the matrix M and vector v given by

$$M = \begin{pmatrix} 1 & 5 \\ 2 & 4 \end{pmatrix}, \quad v = \begin{pmatrix} 1 \\ 1 \end{pmatrix}.$$

Then

$$Mv = \begin{pmatrix} 1 & 5 \\ 2 & 4 \end{pmatrix} \begin{pmatrix} 1 \\ 1 \end{pmatrix} = \begin{pmatrix} 6 \\ 6 \end{pmatrix} = 6 \begin{pmatrix} 1 \\ 1 \end{pmatrix} = 6v.$$

Therefore 6 is an eigenvalue of M, with eigenvector v.

A.4.1   *Calculating Eigenvalues and Eigenvectors*

Given a matrix M, how do we find its eigenvalues and eigenvectors? Equation (A.4) can be written as follows:

$$Mv = \lambda v \;\Rightarrow\; Mv - \lambda I v = 0 \;\Rightarrow\; (M - \lambda I)v = 0, \tag{A.5}$$

where I is the identity matrix. Now, if the matrix $M - \lambda I$ is invertible, we can multiply by its inverse and conclude that v is the zero vector. But this is a contradiction, since in the definition of an eigenvector, v is required to be non-zero. Hence the matrix $M - \lambda I$ must be a singular matrix, with no inverse, which implies that its determinant is zero. So the eigenvalues are found from the equation

$$|M - \lambda I| = 0, \tag{A.6}$$

where the vertical bars indicate the determinant of the matrix. For an $n \times n$ matrix, this is a polynomial of degree n for λ, so in general there are n eigenvalues. Some of the eigenvalues may be complex, and in some cases there may be multiple roots, so there may not be n distinct eigenvalues. Equation (A.6) is called the *characteristic equation* of the matrix M.

Having found the eigenvalues, (A.5) is used to find the eigenvector corresponding to each eigenvalue. An important property of eigenvectors is that if v is an eigenvector, so is cv for any non-zero constant c. So eigenvectors are not uniquely defined, and when finding an eigenvector it is always necessary to make a choice for one of the components of the vector, as shown in the following Example.

Example A.3 Find the eigenvalues and eigenvectors of the matrix

$$M = \begin{pmatrix} 1 & 5 \\ 2 & 4 \end{pmatrix}.$$

The eigenvalues are found from Eq. (A.6),

$$\begin{vmatrix} 1 - \lambda & 5 \\ 2 & 4 - \lambda \end{vmatrix} = 0 \Rightarrow (1 - \lambda)(4 - \lambda) - 10 = 0$$

$$\Rightarrow \lambda^2 - 5\lambda - 6 = 0$$

$$\Rightarrow (\lambda - 6)(\lambda + 1) = 0,$$

so the eigenvalues are $\lambda = 6$ and $\lambda = -1$.

To find the eigenvectors, consider each eigenvalue separately, and try to find a non-zero vector v that obeys the equation $(M - \lambda I)v = 0$. For the first eigenvalue $\lambda = 6$ this equation is

$$\begin{pmatrix} -5 & 5 \\ 2 & -2 \end{pmatrix} \begin{pmatrix} v_1 \\ v_2 \end{pmatrix} = \begin{pmatrix} 0 \\ 0 \end{pmatrix},$$

where v_1 and v_2 are the components of the vector v. Multiplying this out gives two equations for v_1 and v_2, but these equations are duplicates—they both just tell us that $v_1 = v_2$. Therefore we must make a choice of one of the components, for example $v_1 = 1$, leading to $v_2 = 1$.

For the second eigenvalue, $\lambda = -1$, the eigenvector equation is

$$\begin{pmatrix} 2 & 5 \\ 2 & 5 \end{pmatrix} \begin{pmatrix} v_1 \\ v_2 \end{pmatrix} = \begin{pmatrix} 0 \\ 0 \end{pmatrix}.$$

Again, there is only really one equation, $2v_1 + 5v_2 = 0$, for the two unknowns v_1 and v_2, so we have to make a choice. If this doesn't happen, something must have gone wrong in the calculation! We can choose to write the eigenvector for $\lambda = -1$ as

$$v = \begin{pmatrix} 1 \\ -2/5 \end{pmatrix} \quad \text{or} \quad v = \begin{pmatrix} -5/2 \\ 1 \end{pmatrix} \quad \text{or} \quad v = \begin{pmatrix} 5 \\ -2 \end{pmatrix} \quad \text{or} \quad v = \begin{pmatrix} -5 \\ 2 \end{pmatrix},$$

for example.

Choosing the last option, the eigenvalues and corresponding eigenvectors are

$$\lambda = 6, \ \begin{pmatrix} 1 \\ 1 \end{pmatrix} \quad \text{and} \quad \lambda = -1, \ \begin{pmatrix} -5 \\ 2 \end{pmatrix}.$$

A.4.2   Properties of Eigenvalues and Eigenvectors

This section lists some important properties of eigenvalues and eigenvectors of matrices. Although this book is mainly concerned with 2×2 matrices, these results hold for $n \times n$ matrices.

1. The sum of the eigenvalues of M is the trace of M, that is, the sum of the elements on the diagonal of the matrix.

$$T = \text{Trace}(M) = \sum_{j=1}^{n} M_{jj}.$$

2. The product of the eigenvalues of M is the determinant of M.
3. A consequence of the two properties above is that for a 2×2 matrix, the characteristic polynomial is

$$\lambda^2 - T\lambda + D = 0,$$

   where T is the trace of the matrix and D is its determinant.
4. The eigenvalues of a real matrix can be complex, for example

$$\begin{pmatrix} 0 & -1 \\ 1 & 0 \end{pmatrix} \text{ has eigenvalues } \lambda = i, \lambda = -i.$$

   The eigenvalues are complex conjugates of each other. The eigenvectors are also complex conjugates.
5. Usually an $n \times n$ matrix has n distinct eigenvalues and n linearly independent eigenvectors. But if the characteristic polynomial has multiple roots, there may not be n independent eigenvectors. For example

$$\begin{pmatrix} 3 & -1 \\ 1 & 1 \end{pmatrix} \text{ has only one eigenvector, } v = \begin{pmatrix} 1 \\ 1 \end{pmatrix},$$

   (apart from, as always, scalar multiples of this vector), with eigenvalue 2.
6. If M is *symmetric*, meaning that $M_{ij} = M_{ji}$, then the eigenvalues of M are real and its eigenvectors are orthogonal to each other.

7. If a matrix is triangular, meaning that either $M_{ij} = 0$ for $i < j$ or $M_{ij} = 0$ for $i > j$, then the eigenvalues are just the numbers on the diagonal of M. For example, the eigenvalues of

$$M = \begin{pmatrix} 3 & 1 & 5 \\ 0 & 4 & 9 \\ 0 & 0 & 1 \end{pmatrix} \quad \text{are } \lambda = 3, \lambda = 4, \lambda = 1.$$

A.5   Partial Derivatives

Suppose that $f(x, y)$ is a function of two independent variables x and y. Then the *partial derivative* of f with respect to x is defined as the derivative with respect to x while y is kept constant,

$$\frac{\partial f}{\partial x} = \lim_{h \to 0} \frac{f(x + h, y) - f(x, y)}{h},$$

as long as this limit exists, and the partial derivative with respect to y is defined similarly. The special symbol ∂ is used to distinguish between partial and ordinary derivatives. A shorthand notation often used for the partial derivatives is f_x, f_y. Second derivatives f_{xx} and f_{yy} are defined in a similar way, and differentiating with respect to x and then with respect to y or vice versa gives the cross-derivative $f_{xy} = f_{yx}$.

If small changes Δx and Δy are made to x and y, the corresponding small change to f is approximately

$$\Delta f \approx \frac{\partial f}{\partial x} \Delta x + \frac{\partial f}{\partial y} \Delta y. \tag{A.7}$$

This formula is in fact the beginning of the *Taylor series* for a function of two variables,

$$f(x + \Delta x, y + \Delta y) = f(x, y) + \frac{\partial f}{\partial x} \Delta x + \frac{\partial f}{\partial y} \Delta y$$

$$+ \frac{1}{2} \left(\frac{\partial^2 f}{\partial x^2} \Delta x^2 + 2 \frac{\partial^2 f}{\partial x \partial y} \Delta x \Delta y + \frac{\partial^2 f}{\partial y^2} \Delta y^2 \right) + O(3).$$

Closely related to (A.7) is the *chain rule* for functions of two variables. If x and y are functions of u and v, then

$$\frac{\partial f}{\partial u} = \frac{\partial f}{\partial x} \frac{\partial x}{\partial u} + \frac{\partial f}{\partial y} \frac{\partial y}{\partial u}, \tag{A.8}$$

and similarly, using the shorthand notation, $f_v = f_x x_v + f_y y_v$.

All of these definitions and properties can be extended to functions of three or more variables.

A.6   Vector Calculus

Suppose that $f(x, y, z)$ is a scalar function of x, y and z, corresponding to Cartesian coordinates in three-dimensional space. This function is often described as a *scalar field*. Physical examples include temperature or density. Similarly, a *vector field* is a vector-valued function of position, $\mathbf{u}(x, y, z)$, such as wind velocity or magnetic field.

There are three important ways in which scalar and vector fields can be differentiated, leading to another scalar or vector field. These three quantities, grad, div and curl, are all independent of any coordinate system, in other words, they are properties of the scalar or vector field itself.

The *gradient* of a scalar field is a vector field, grad f $= \nabla f$, with components consisting of the three partial derivatives of f:

$$\nabla f = \left(\frac{\partial f}{\partial x}, \frac{\partial f}{\partial y}, \frac{\partial f}{\partial z} \right). \tag{A.9}$$

The vector ∇f is normal (perpendicular) to the surface on which $f = $ constant, pointing in the direction of increasing f. Its magnitude, $|\nabla f|$, is the rate of change of f in this spatial direction.

The *divergence* of a vector field $\mathbf{u} = (u_1, u_2, u_3)$ is a scalar field written as div $\mathbf{u}$ or $\nabla \cdot \mathbf{u}$, defined by

$$\nabla \cdot \mathbf{u} = \frac{\partial u_1}{\partial x} + \frac{\partial u_2}{\partial y} + \frac{\partial u_3}{\partial z}. \tag{A.10}$$

The divergence can be thought of as a measure of the expansion associated with the vector field. If $\mathbf{u}$ is the velocity vector of an incompressible fluid, then its divergence must be zero. The notation $\nabla \cdot \mathbf{u}$ is used because the divergence can be thought of as the dot product of the vector differential operator

$$\nabla = \left(\frac{\partial}{\partial x}, \frac{\partial}{\partial y}, \frac{\partial}{\partial z} \right)$$

and the vector $\mathbf{u}$.

The third type of derivative is the *curl* of a vector field $\mathbf{u}$, $\nabla \times \mathbf{u}$, which is the cross product of the differential operator ∇ and $\mathbf{u}$. In component form,

$$\nabla \times \mathbf{u} = \left(\frac{\partial u_3}{\partial y} - \frac{\partial u_2}{\partial z}, \frac{\partial u_1}{\partial z} - \frac{\partial u_3}{\partial x}, \frac{\partial u_2}{\partial x} - \frac{\partial u_1}{\partial y} \right). \tag{A.11}$$

As its name suggests, the curl is a measure of the rotation rate of the vector field at a point.

Recommended Further Reading

There are many excellent books on differential equations and chaos. The following list includes some of these. The first contains no mathematics, but tells the story of the discoveries of Lorenz, Feigenbaum and others in a readable and entertaining way. The last is probably the most thorough and rigorous.

Gleick, J. (1997) Chaos: Making a New Science. Random House, New York

Acheson, D. (1997) From Calculus to Chaos: An Introduction to Dynamics. Oxford University Press

Strogatz, S.H. (2018) Nonlinear Dynamics and Chaos: With Applications to Physics, Biology, Chemistry, and Engineering. CRC Press

Blanchard, P., Devaney, R.L., Hall, G. (1995) Differential Equations

Drazin, P.G. (1992) Nonlinear Systems. Cambridge University Press

Robinson, J.C. (2004) An Introduction to Ordinary Differential Equations. Cambridge University Press

Glendinning, P. (1994) Stability, Instability and Chaos: An Introduction to the Theory of Nonlinear Differential Equations. Cambridge University Press

Wiggins, S. (2003) Introduction to Applied Nonlinear Dynamical Systems and Chaos. Springer

© The Author(s), under exclusive license to Springer Nature Switzerland AG 2025   231
P. C. Matthews, *Differential Equations, Bifurcations and Chaos*,
Springer Undergraduate Mathematics Series,
https://doi.org/10.1007/978-3-031-99543-9

Index